The Studio-X NY Guide to Liberating New Forms of Conversation

The Studio-X NY Guide to Liberating New Forms of Conversation

edited by Gavin Browning
afterword by Mark Wigley

GSAPP Books
New York

Published by
The Graduate School of Architecture, Planning and Preservation
Columbia University
1172 Amsterdam Ave.
409 Avery Hall
New York, NY 10027

Visit our website at *arch.columbia.edu/publications*

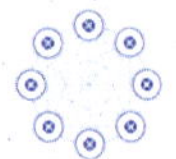

Studio-X Report 002.

Produced through the Office of the Dean, Mark Wigley and
the Office of Print Publications.

Thanks to the Office of the Dean at Columbia GSAPP: Leslie
Bailey, Jessica Braun, David Hinkle, Janet Reyes, Danielle
Smoller and Mark Wigley; to Craig Buckley in the Office of Print
Publications, to MTWTF for their beautiful design, to Tina
Gao and Greta Byrum for their careful readings, and to the
contributors to this book. Thanks also to Nora Arkawi, George
Browning, Margie Browning, Malwina Łys-Dobradin, Leah
Meisterlin, Ben Prosky and William Brian Smith for their
tremendous help, and to all who have helped to make Studio-X
New York what it is today.

All text written by Gavin Browning unless otherwise noted.

Editor: Gavin Browning
Series Editor: Craig Buckley
Graphic Design: MTWTF, NY: Glen Cummings, Aliza Dzik,
Jin Kim, Juan Astasio, Lanny Hoang, Clara Emo-Dambry
Reader: Greta Byrum
Printer: Linco Printing

Library of Congress Publication Data:
1-883584-65-5

The Studio-X New York Guide To Liberating New Forms
of Conversation/edited by Gavin Browning ; afterword by Mark
Wigley. —1st ed.
192 p. 14 x 20.6718 cm.
Includes index.
ISBN 1-883584-65-5
1. Communication in architectural design—New York (State)—
New York. 2. Forums (Discussion and debate)—New York
(State)—New York. 3. Architecture—Study and teaching
(Graduate)—New York (State)—New York. I. Browning, Gavin.
ZII. Columbia University. Graduate School of Architecture,
Planning, and Preservation.
NA2750.S92 2010
720.71'17471—dc22
                    2010028294

This Book Will Set You Free

What does it mean to liberate new forms
of conversation?

This book presents both a method and a manual for
fostering conversation, and a record of a series of new
conversations held at Studio-X New York, a flexible,
high-traffic, work/event space on the 16th floor of 180
Varick Street in Lower Manhattan. Both method
and practice have evolved through trial and error over
the course of the space's inaugural two years.

Run by the Graduate School of Architecture, Planning
and Preservation of Columbia University, Studio-X
New York is dedicated to experimental research
and public programming around urbanism. During
the day (and sometimes late into the night), various
research labs conduct long or short-term projects
in the space. C-Lab, Living Architecture Lab, Network
Architecture Lab, Spatial Information Design Lab,
S.L.U.M. LAB, and Urban Landscape Lab are just some
of the research groups that have used the studio
or call it home. Sometimes, lab work translates into
public programming, and other times it is wholly
unrelated to the dense calendar of events. These two
kinds of work can align or compete. Each requires
a radically different configuration of space and set of
equipment—research is private, and events are
public. However, the two can co-exist, and in fact, can
occur at the same time.

The dynamic setting that is integral to Studio-X
New York is part of what makes the conversations in
this book new. But look further, to the roster of
speakers listed in the *Populate* section of this book.
When a flavor scientist mingles with Austrian inn-
keepers, interaction designers, ecologists, culinary
historians, civil liberties activists, and urban
planners, it's either the most scintillating or awkward
dinner party you've ever attended. Each of these
singular minds has contributed to the success of
the Studio-X New York experiment—bringing
expertise to bear on the built environment and the
myriad physical and social forces that shape it.

Produced in collaboration with MTWTF, this book
mines the growing Studio-X New York archive of
exhibitions, large-scale gatherings, intimate get-
togethers, parties and informal workshops to offer
transferable lessons in creating conversation.
It looks very local, suggesting what to do with what
you have.

—Gavin Browning

Locate.
Studio-X New York is tucked away
behind an unmarked door on
the sixteenth floor of a nondescript
office building in Lower
Manhattan. How do you get there?

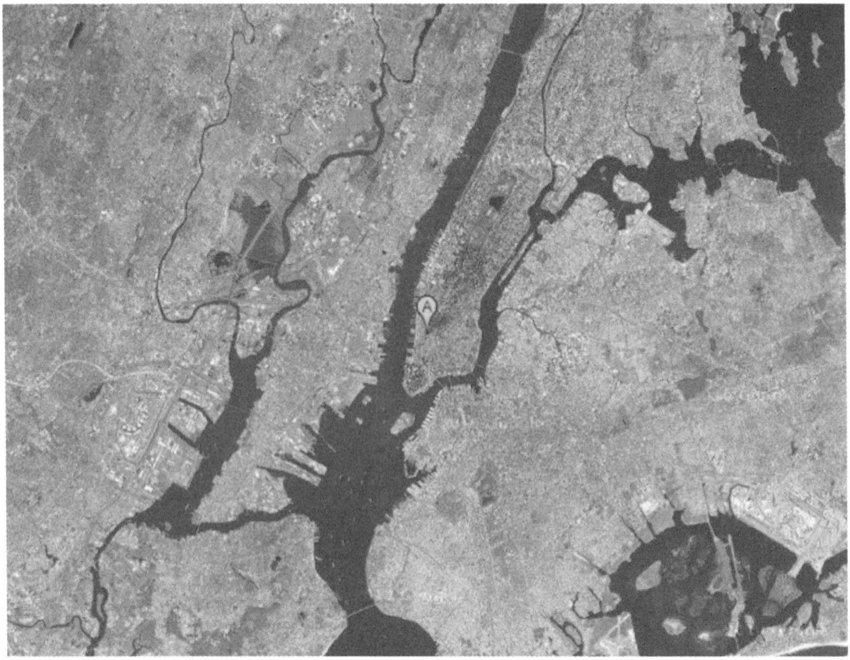

Lower Manhattan

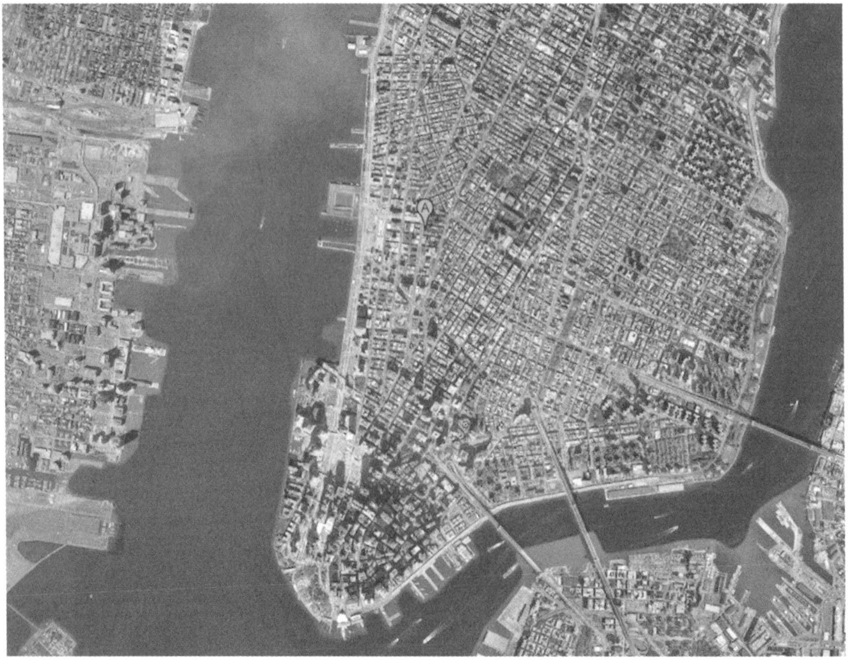

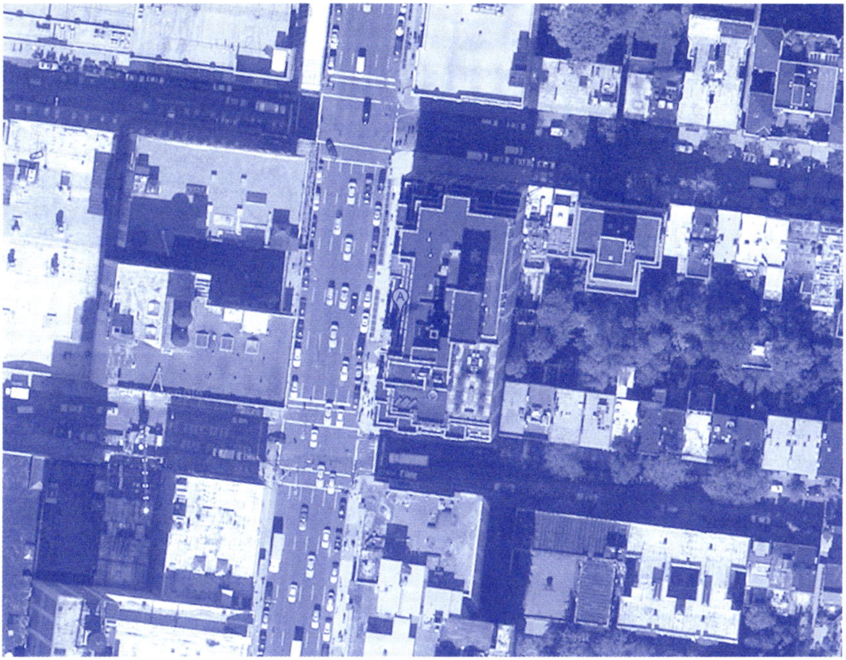

Hudson Square

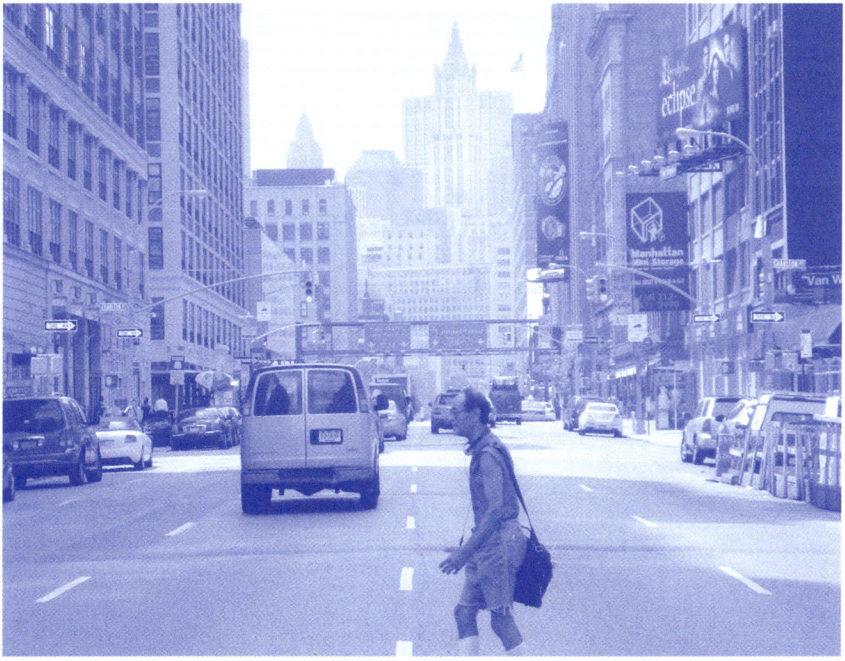

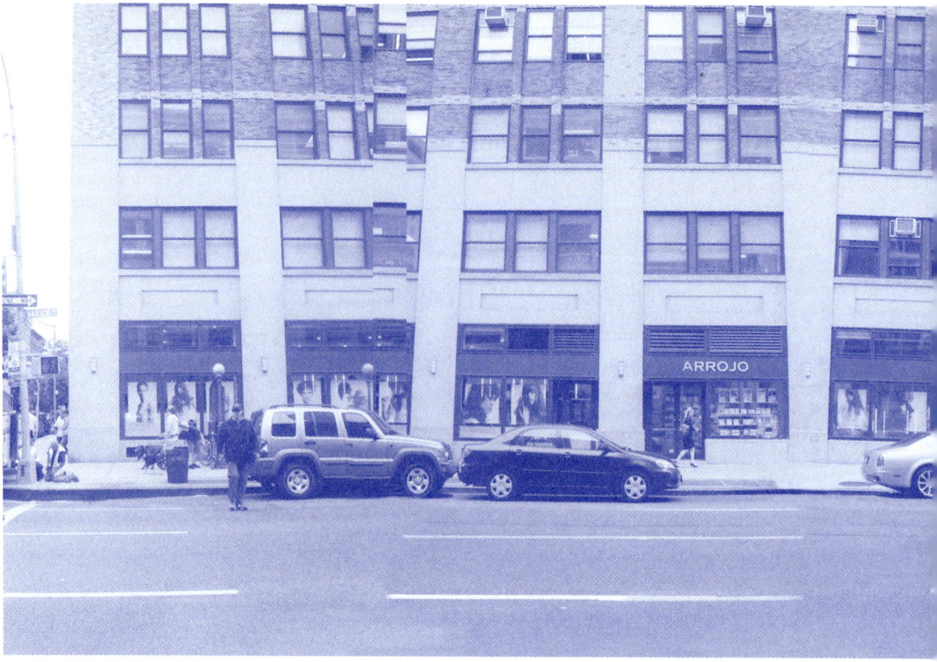
ARROJO

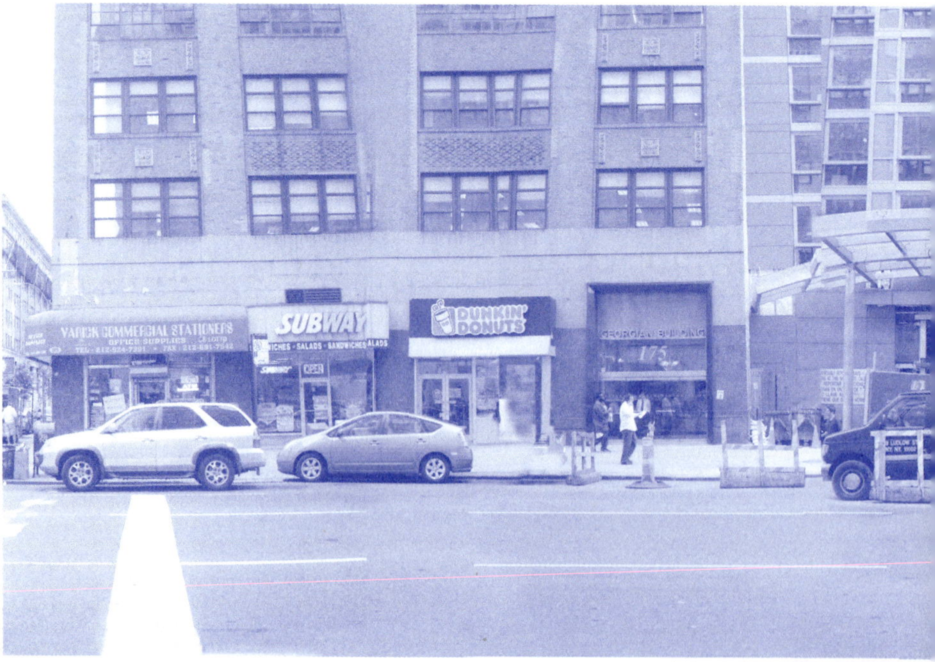
VARICK COMMERCIAL STATIONERS
OFFICE SUPPLIES
SUBWAY
DUNKIN' DONUTS
GEORGIAN BUILDING
175

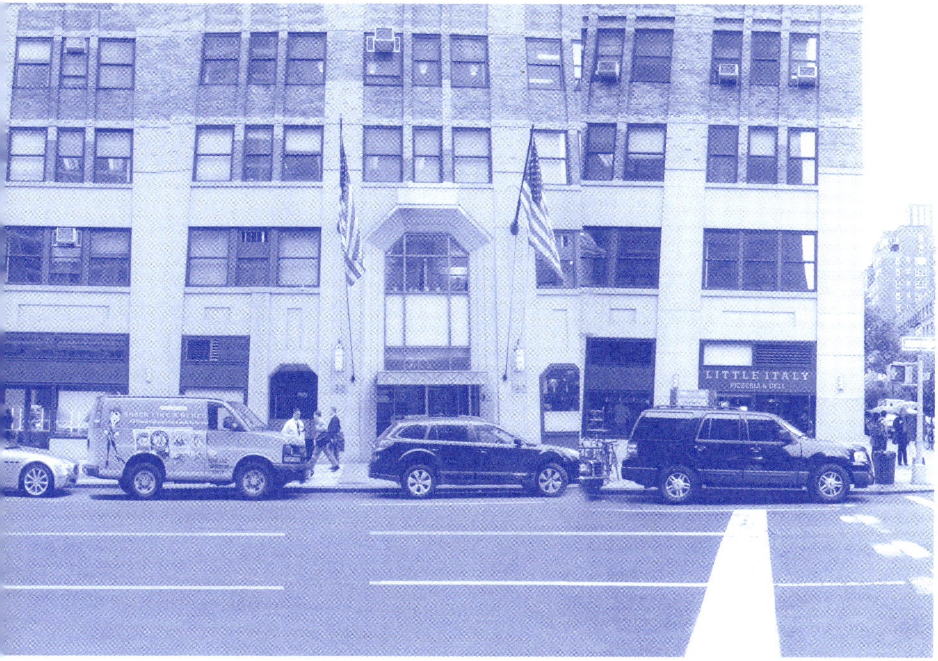

Varick between King and Charlton

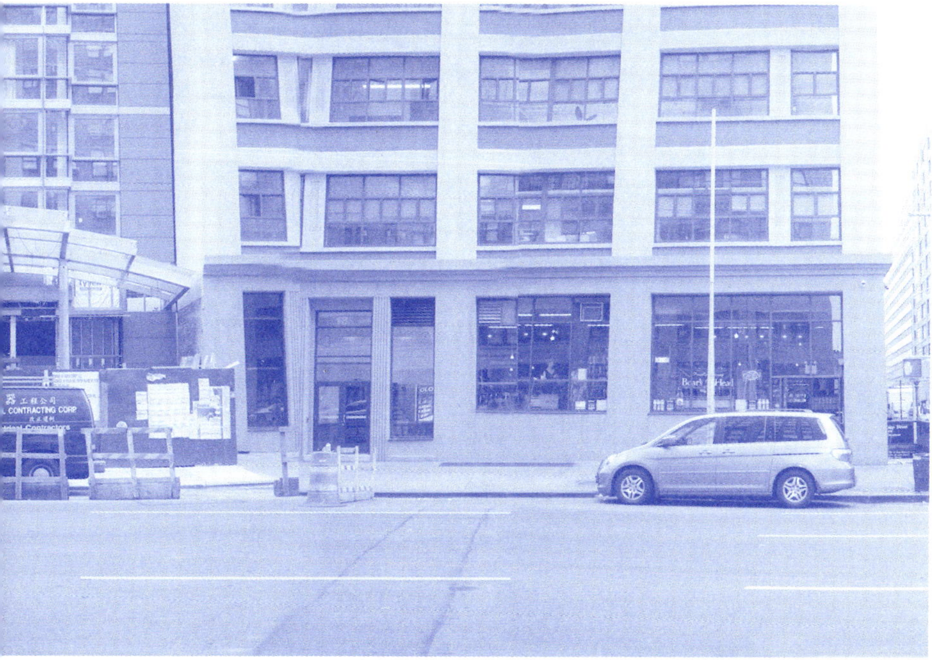

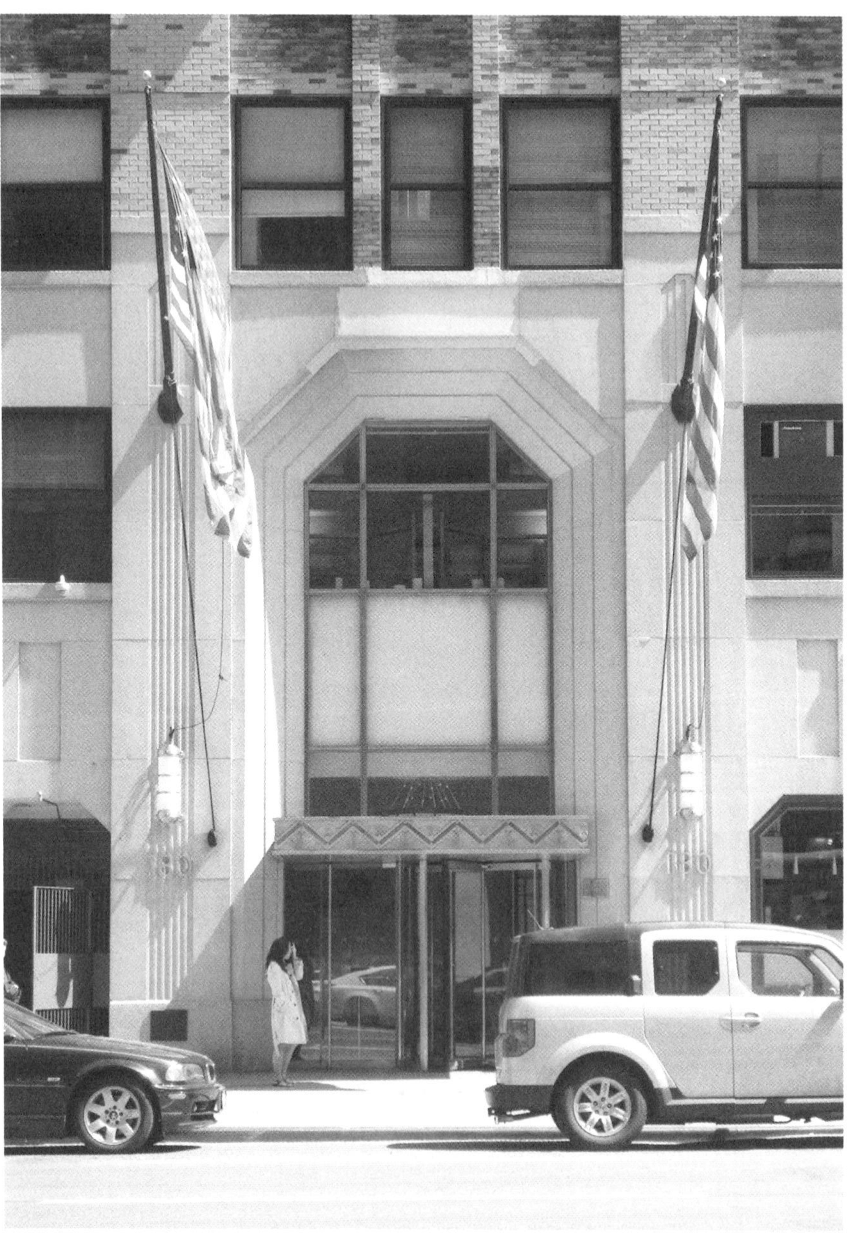

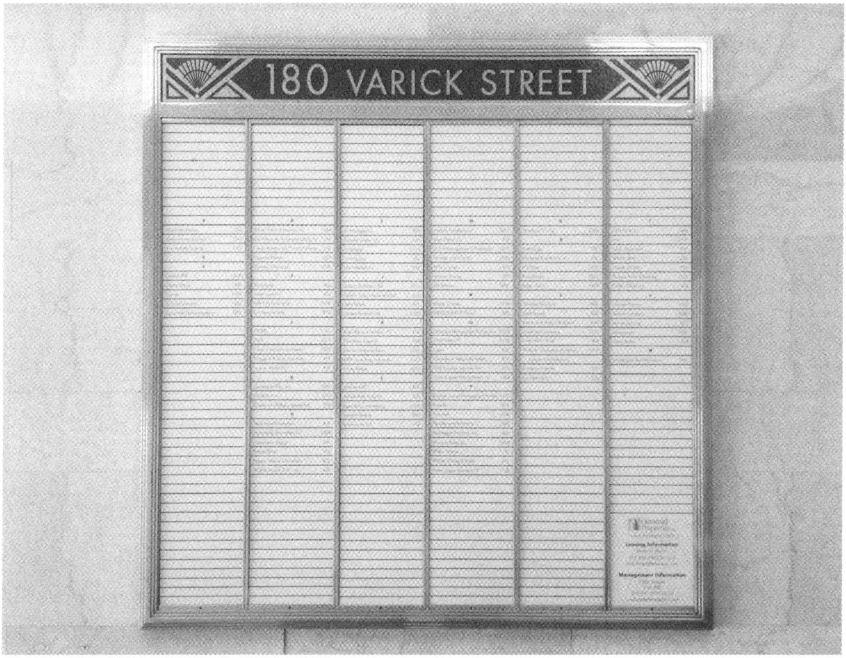

180 Varick Suite 1610

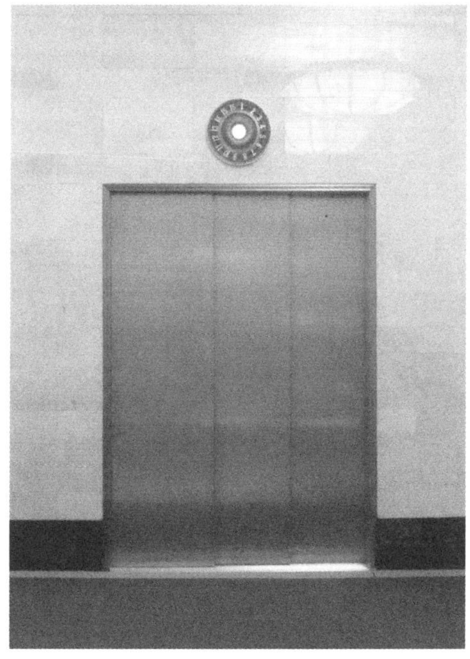

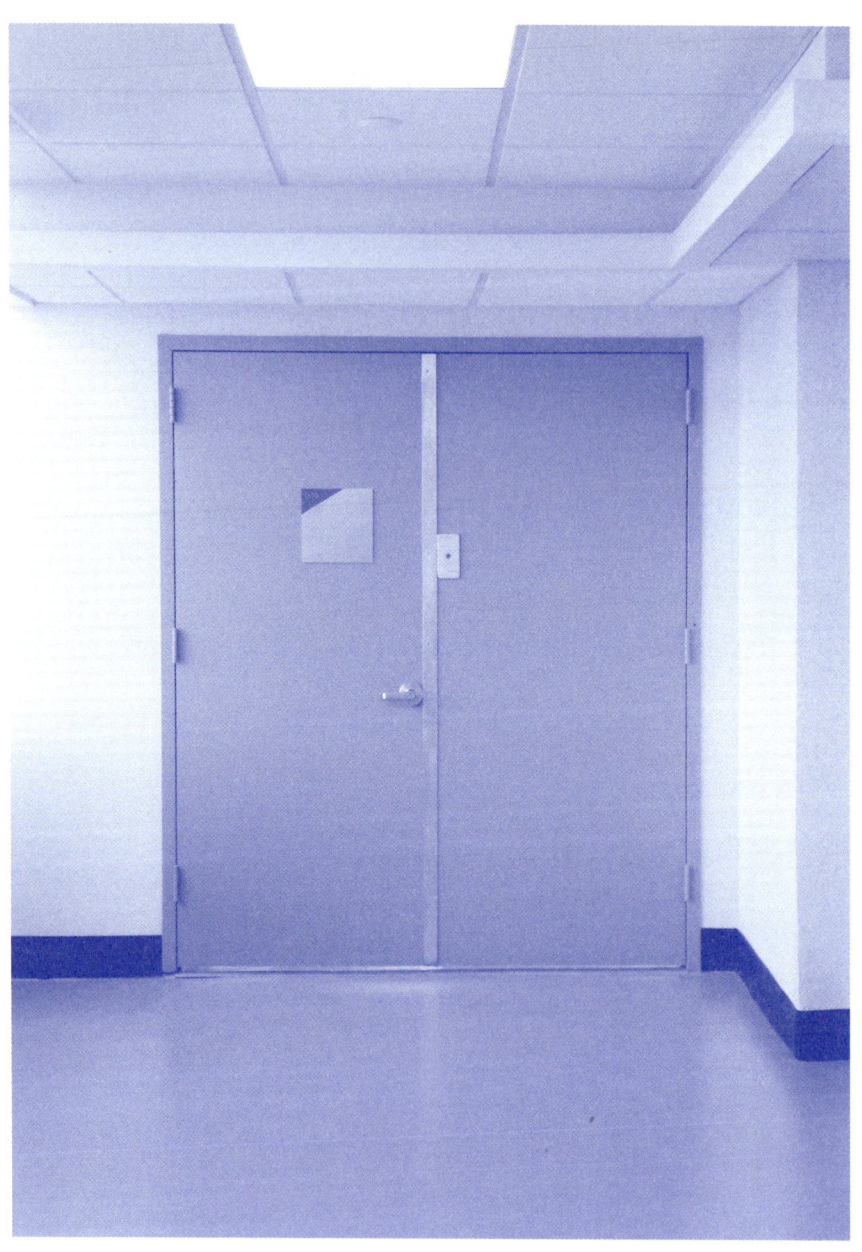

**Enter.
Studio-X New York is used
by different people, for different
purposes, at different times
of day.**

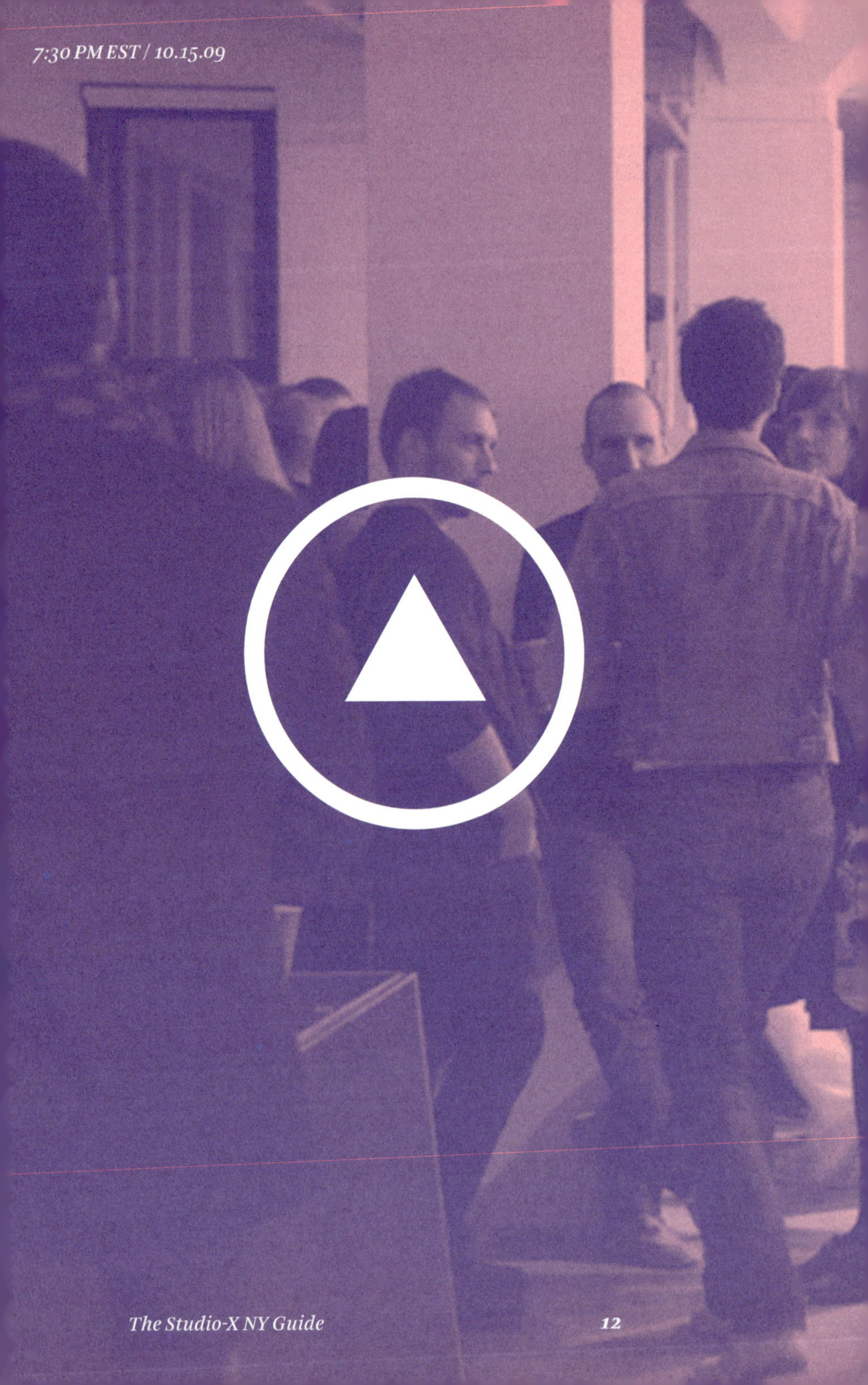

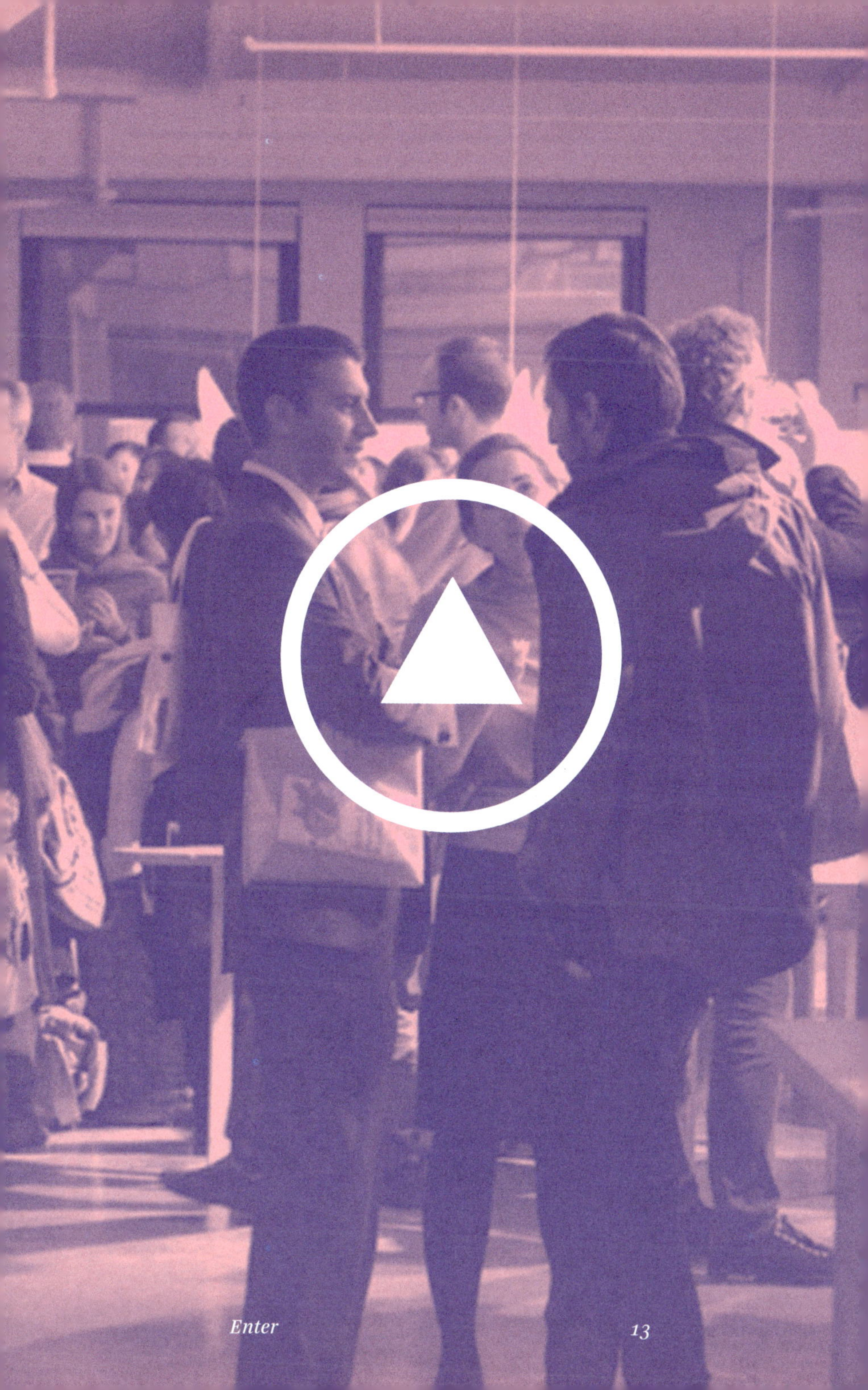

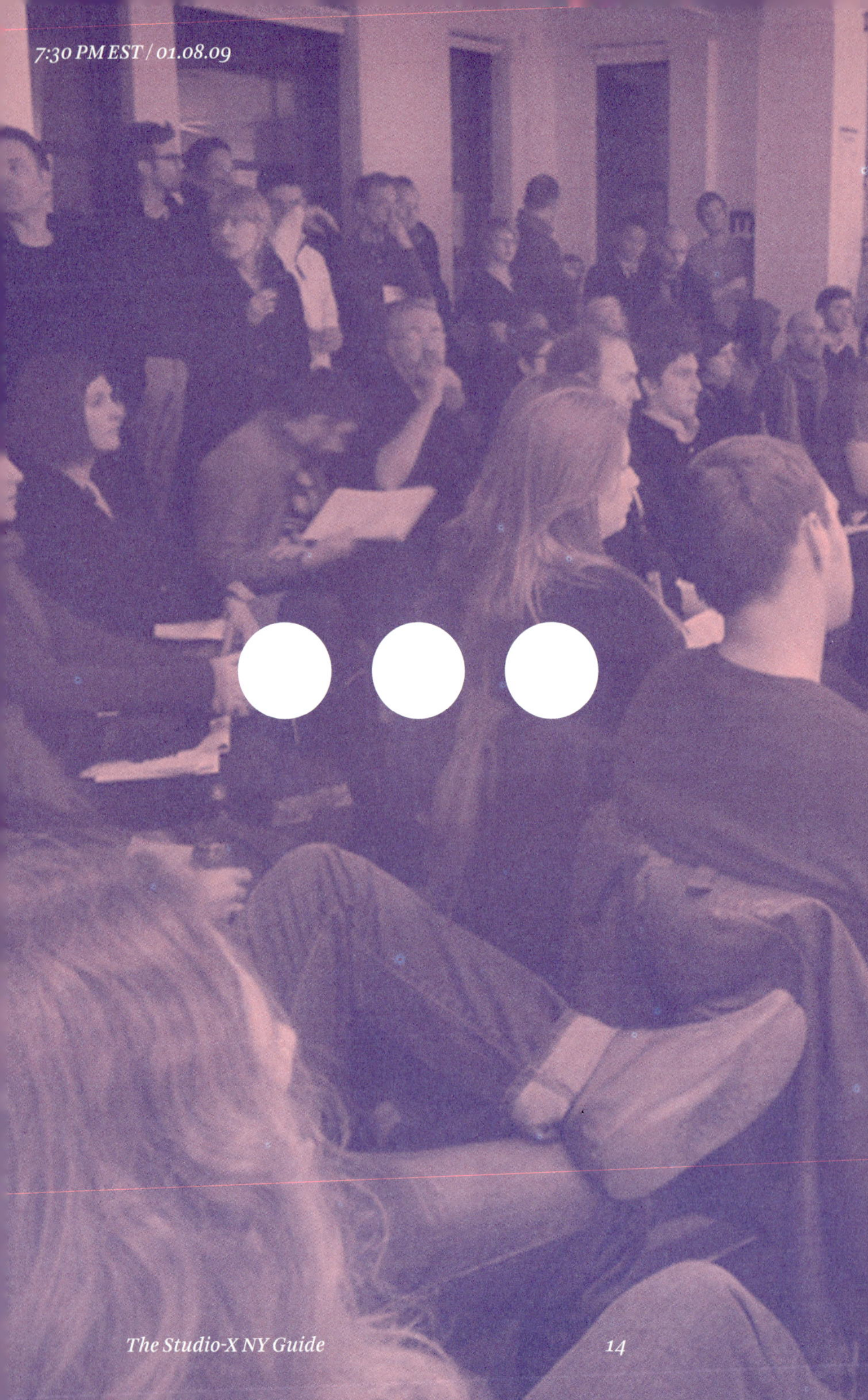
7:30 PM EST / 01.08.09

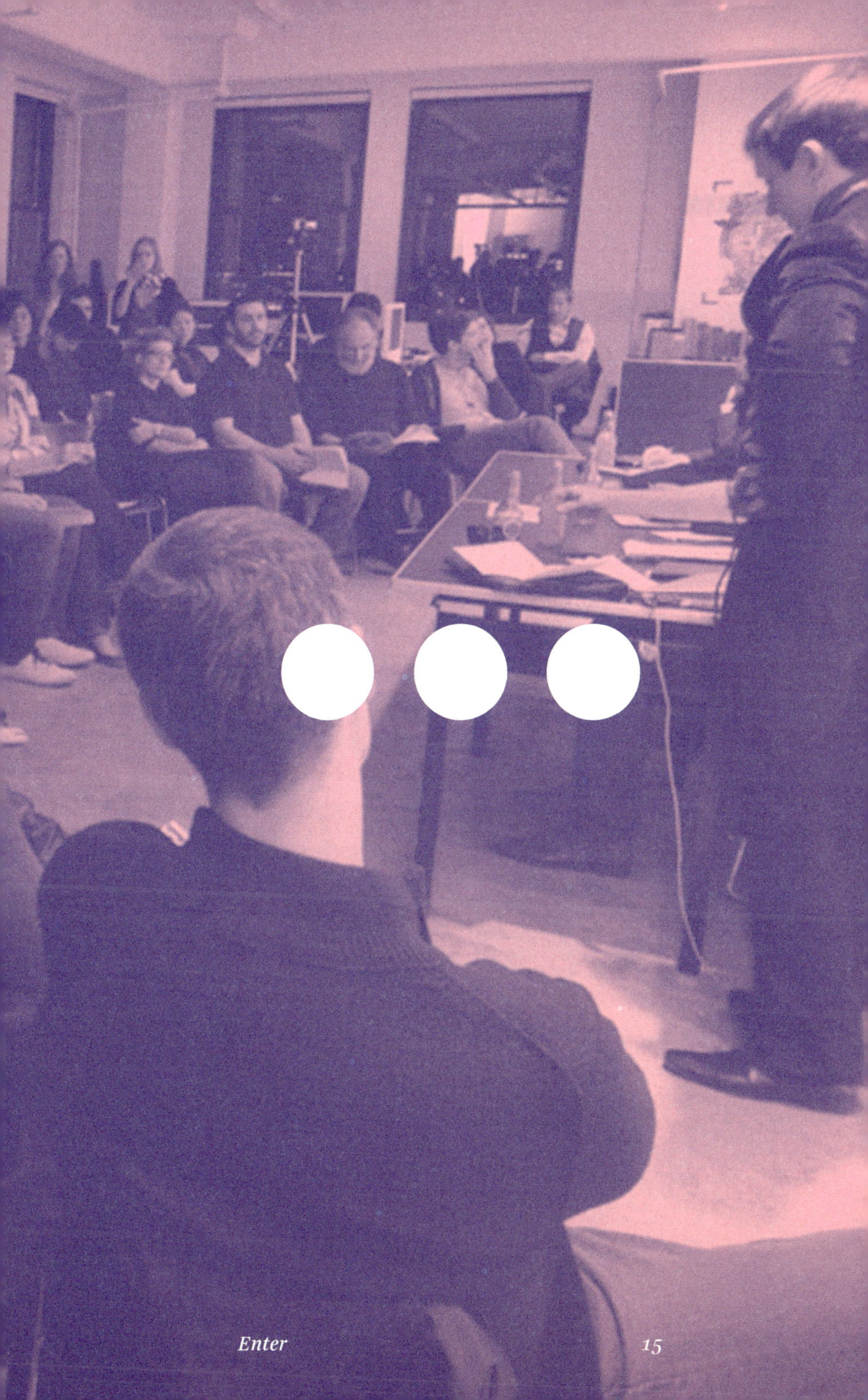

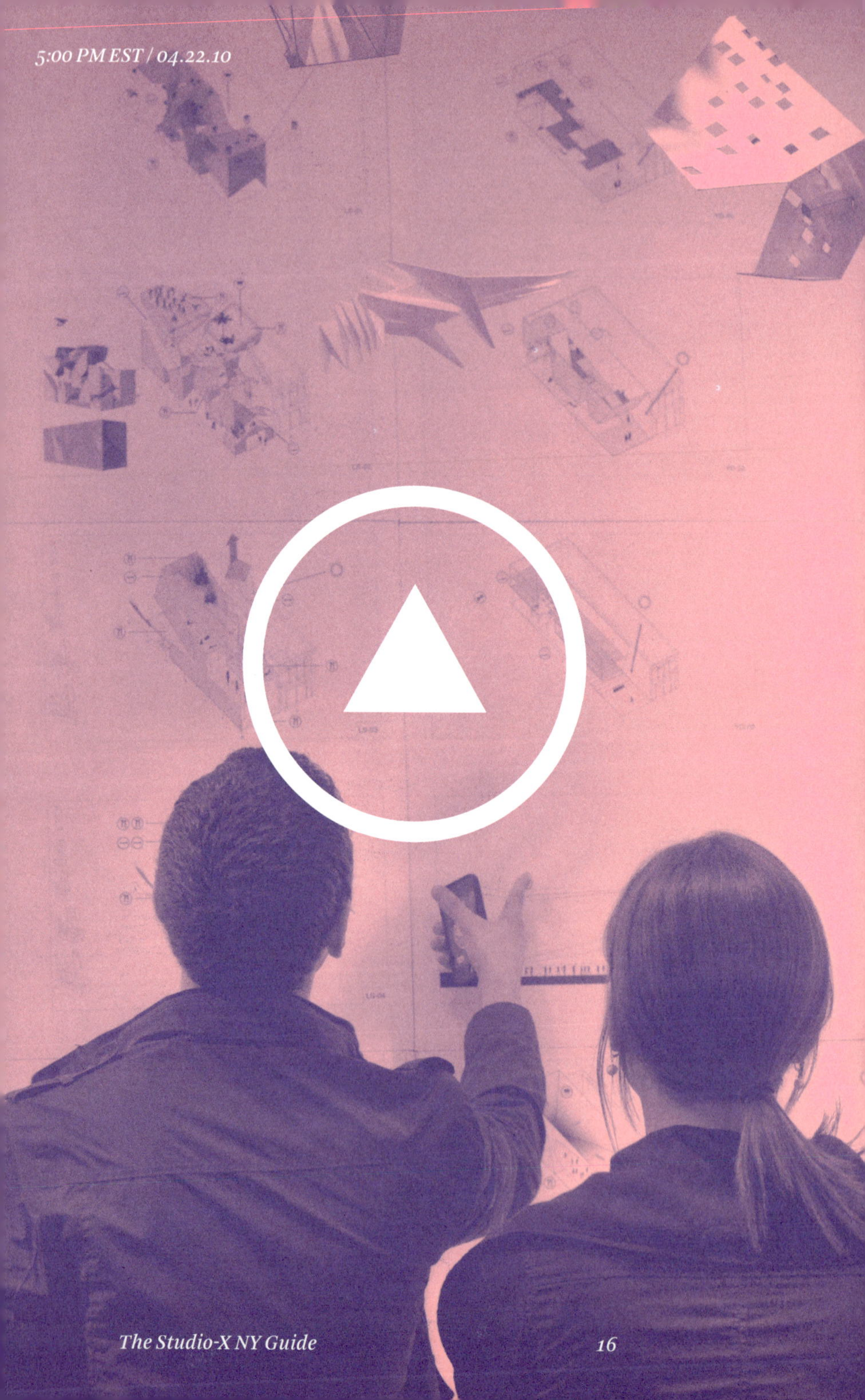
5:00 PM EST / 04.22.10

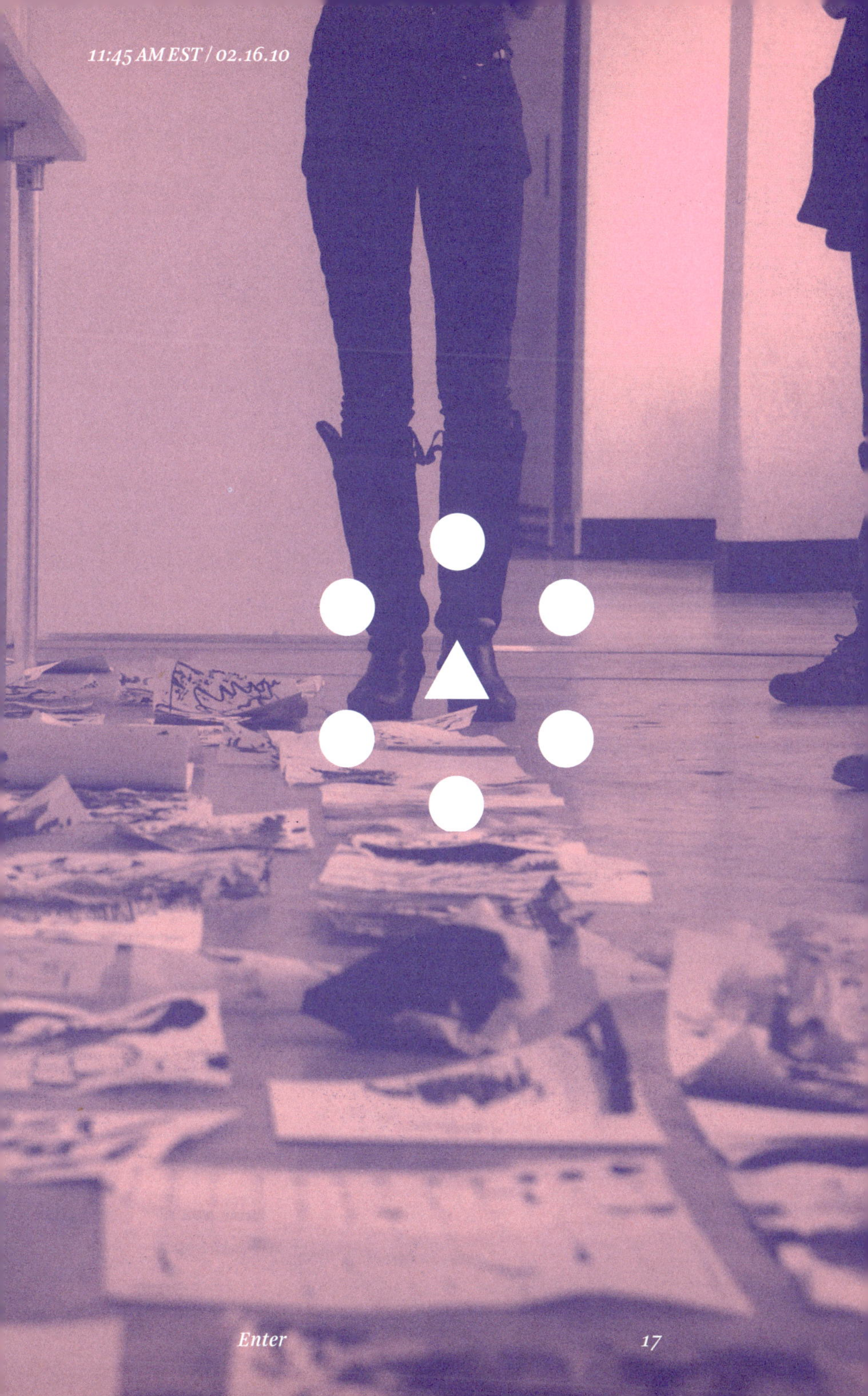
11:45 AM EST / 02.16.10

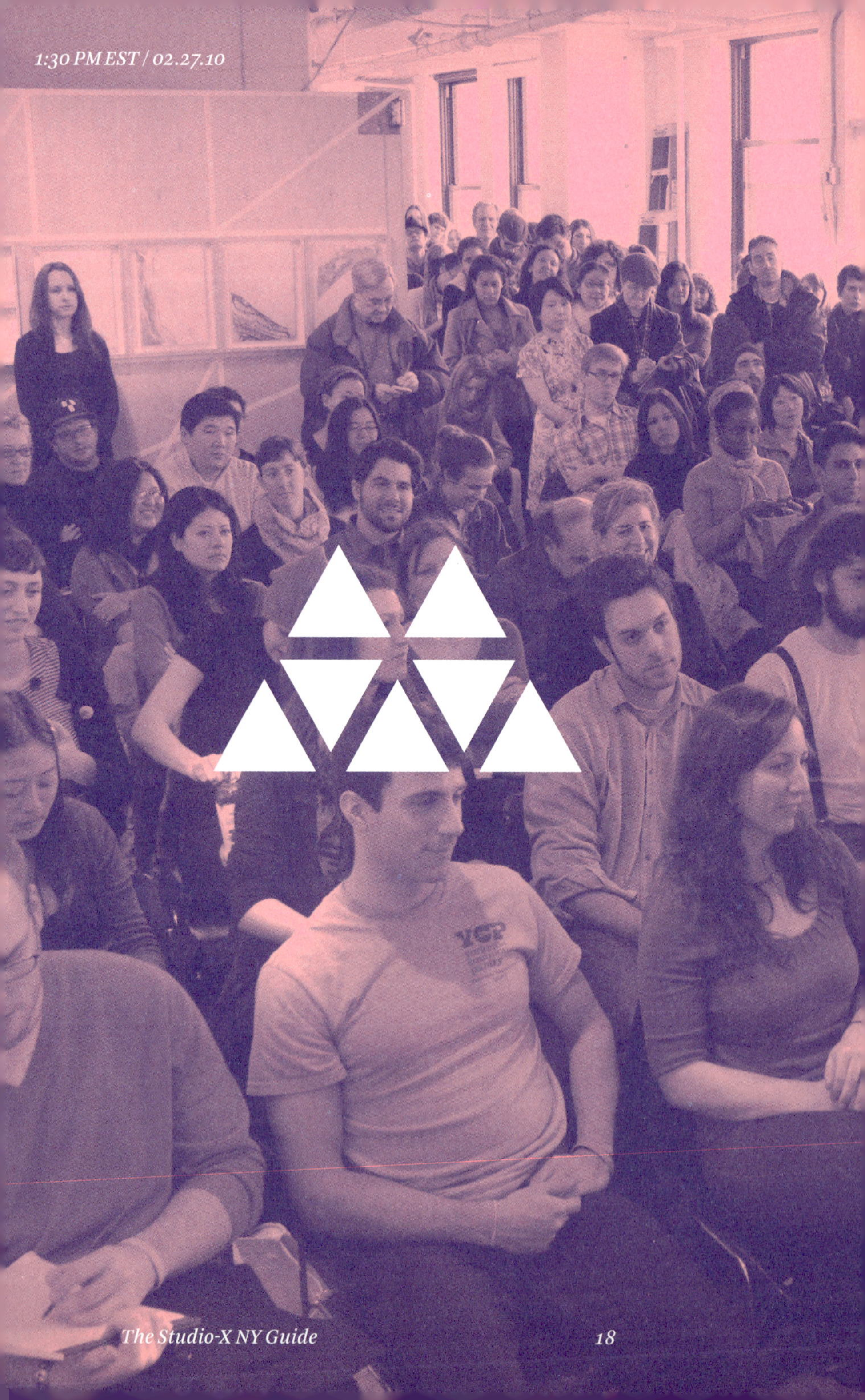

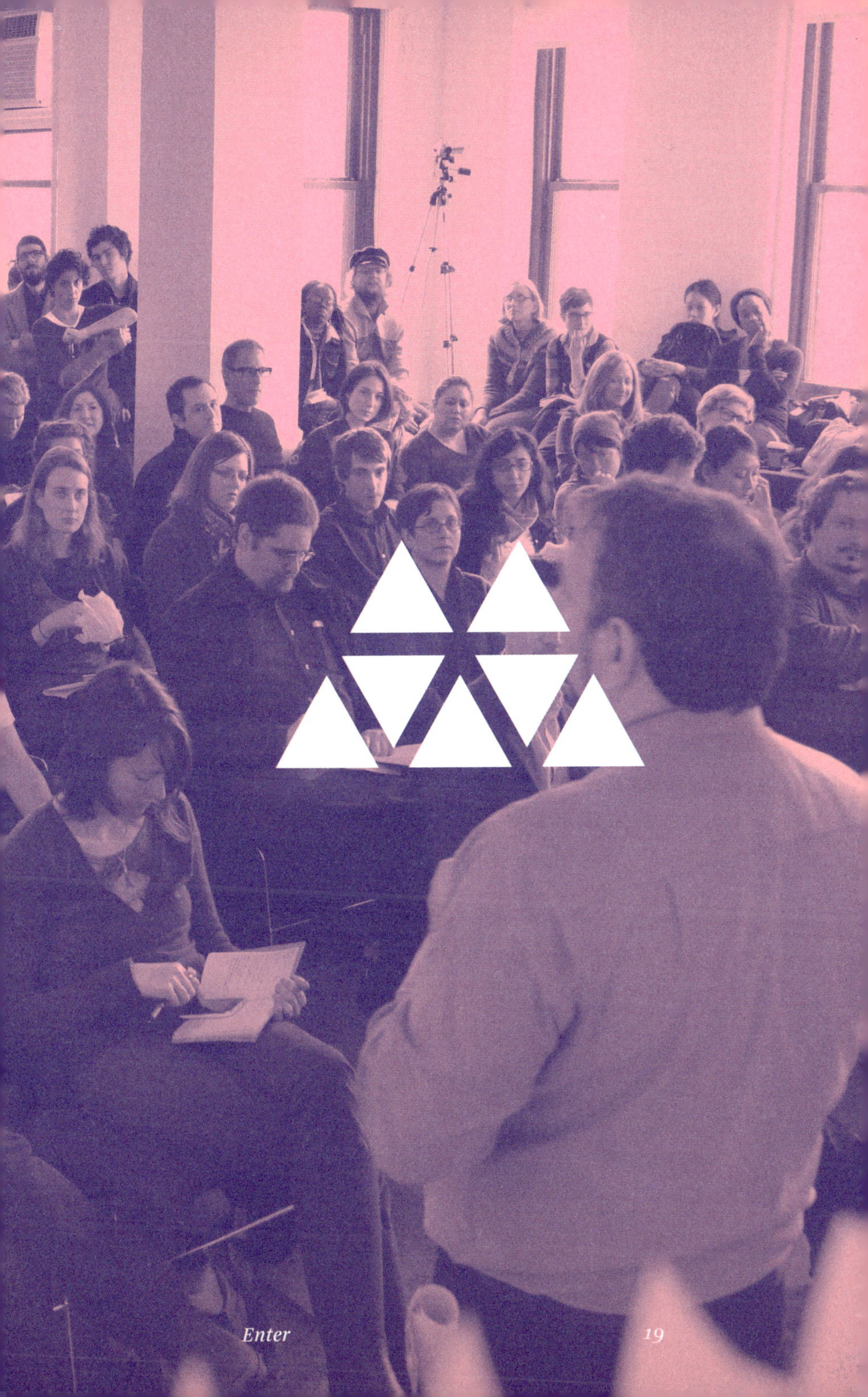

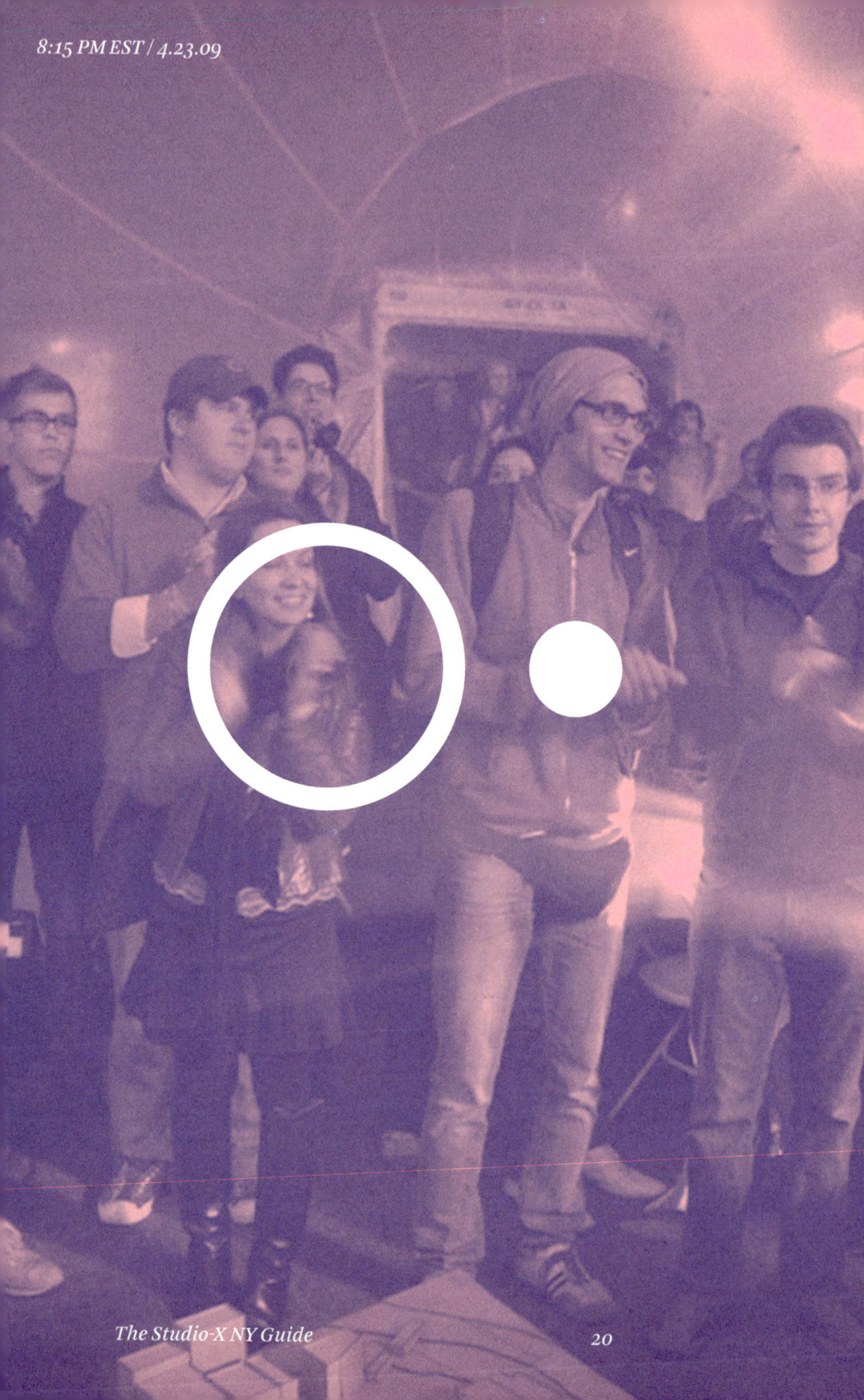
8:15 PM EST / 4.23.09

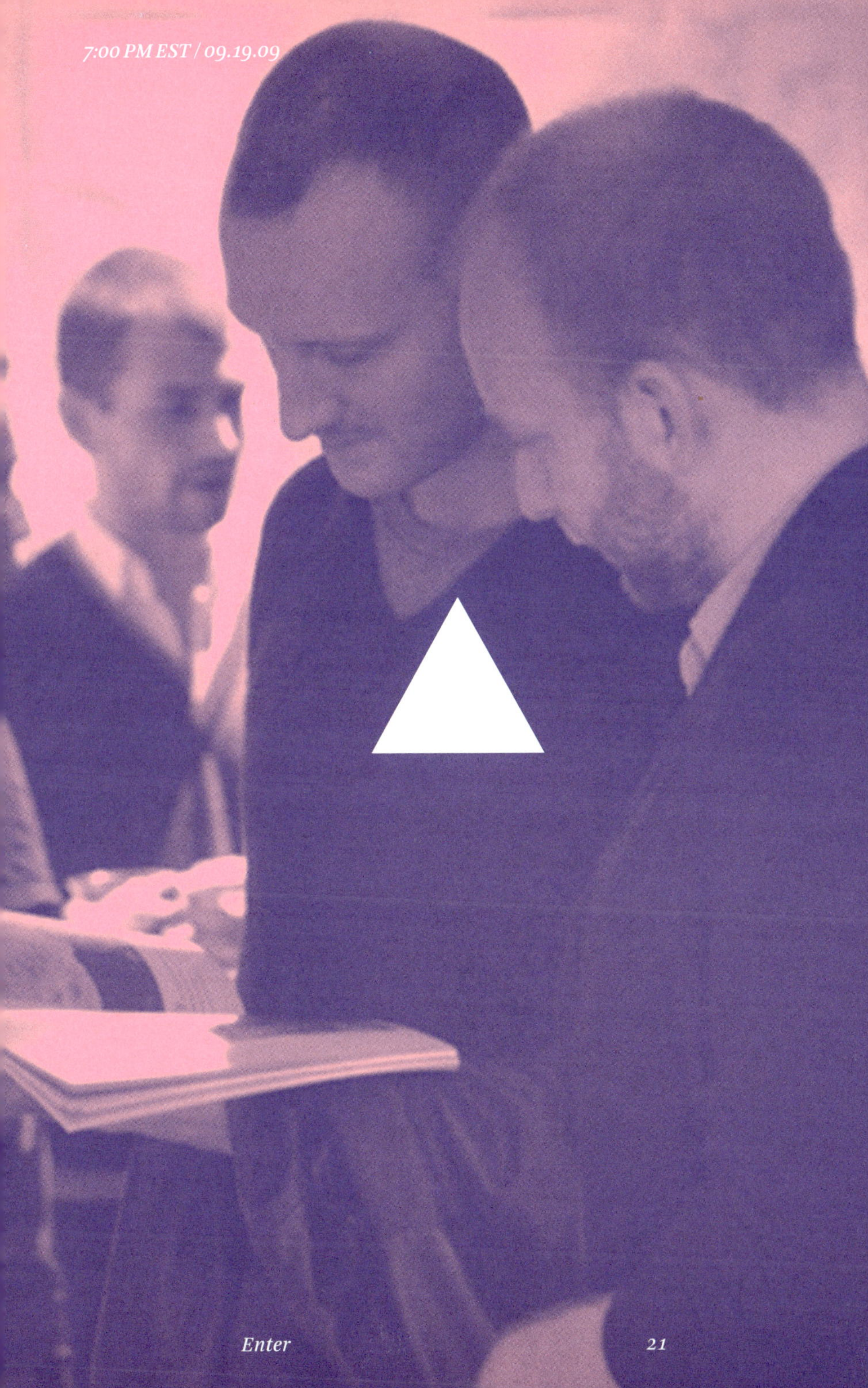

7:00 PM EST / 09.19.09
Enter
21

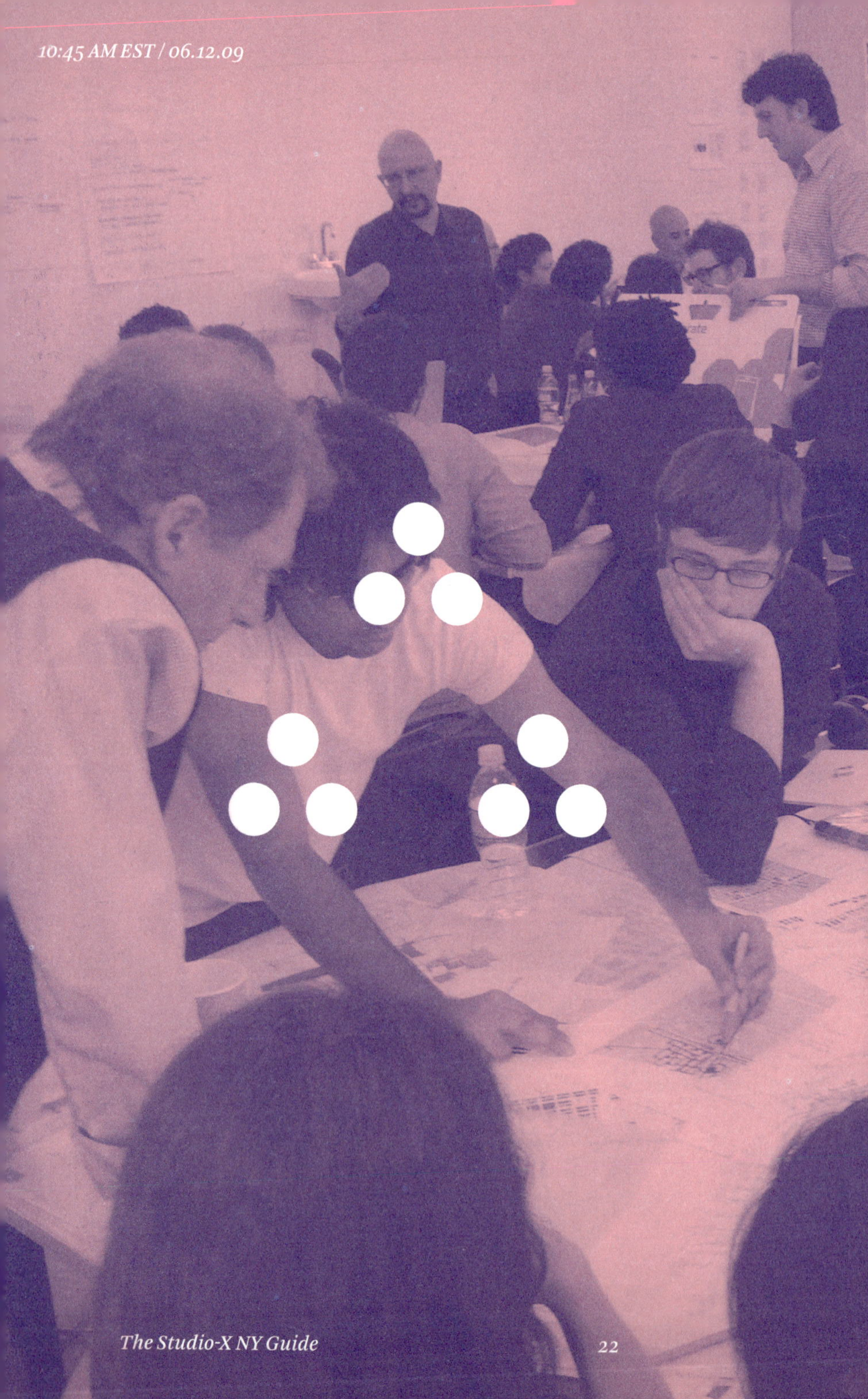
10:45 AM EST / 06.12.09

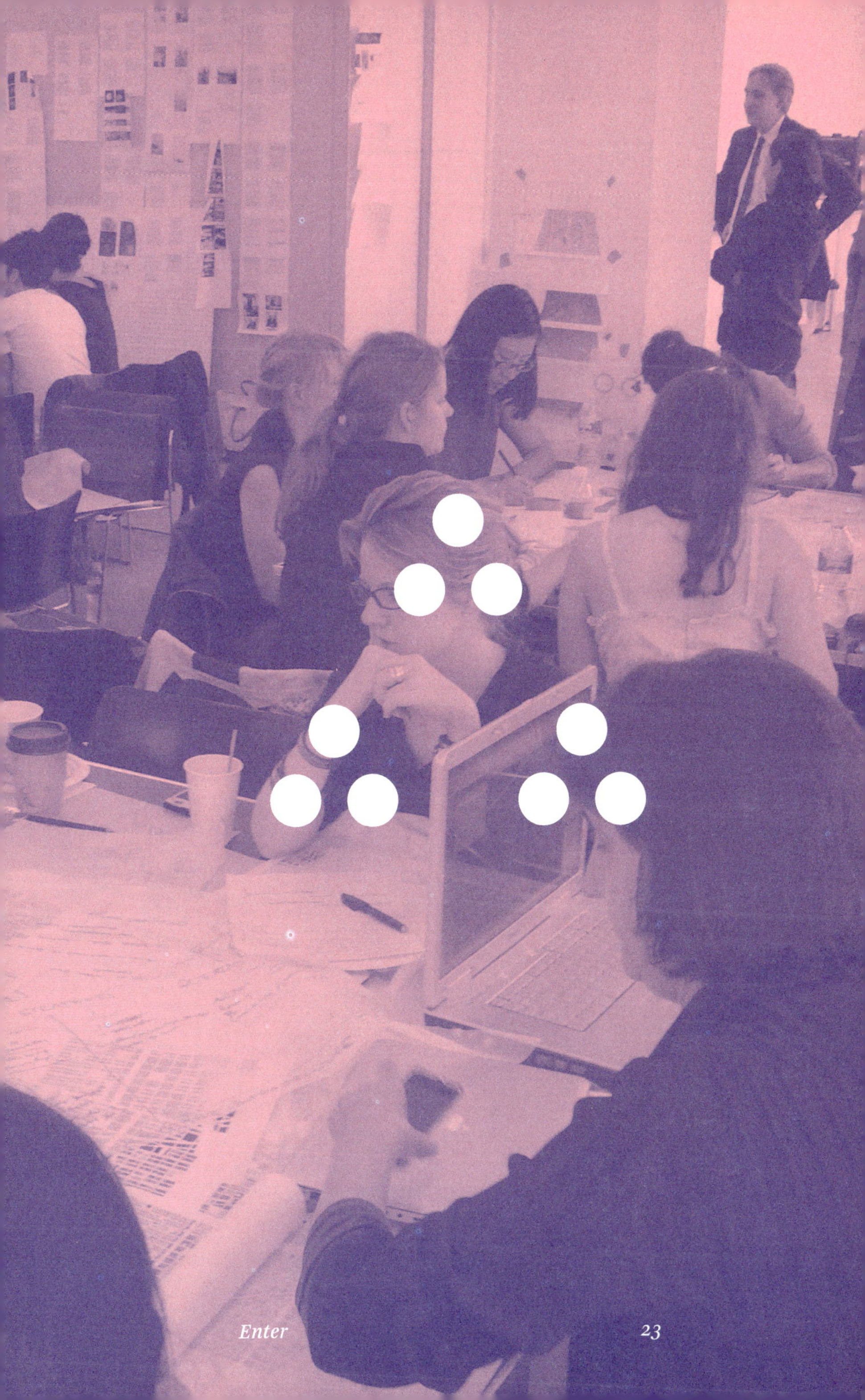

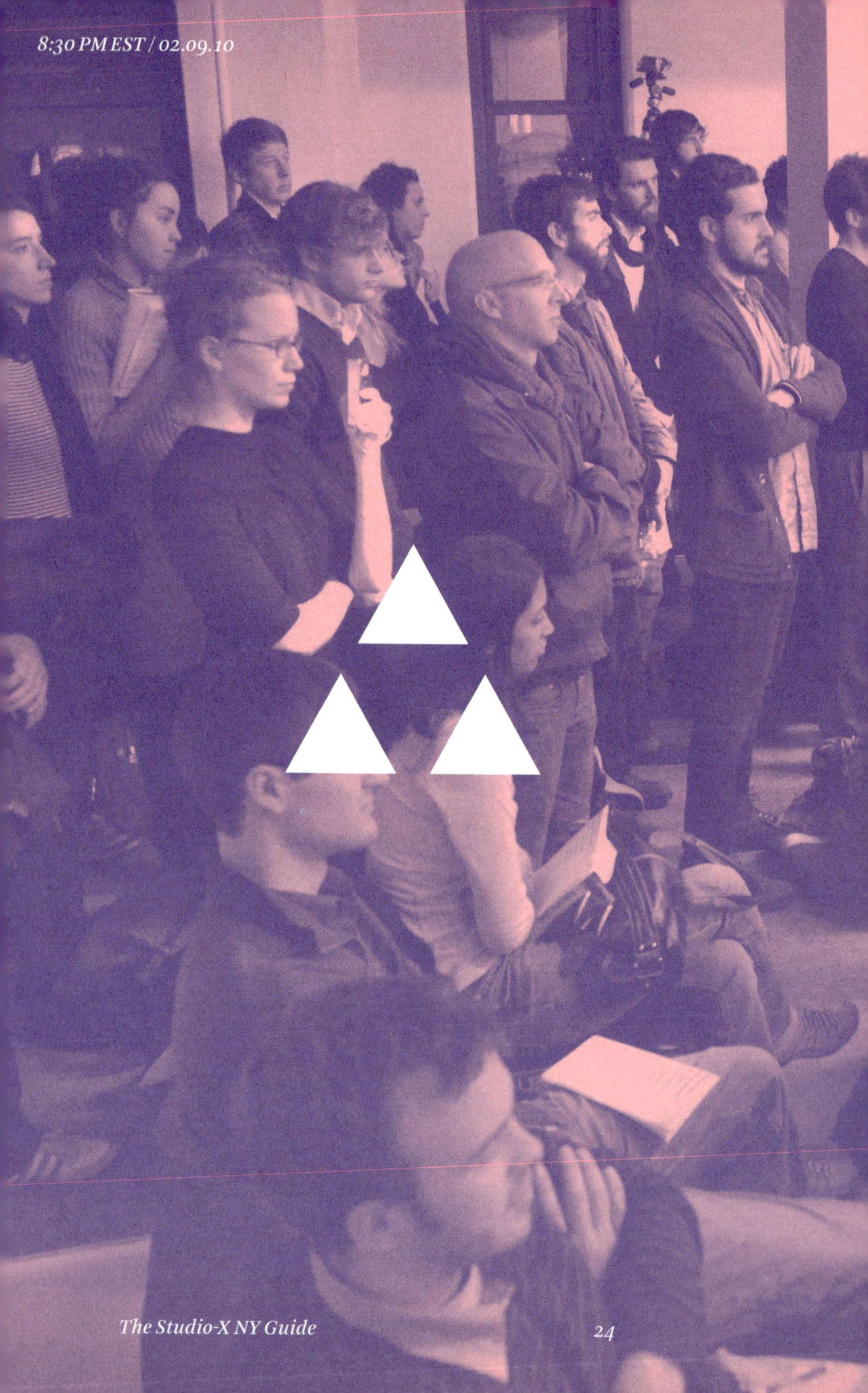
8:30 PM EST / 02.09.10
The Studio-X NY Guide
24

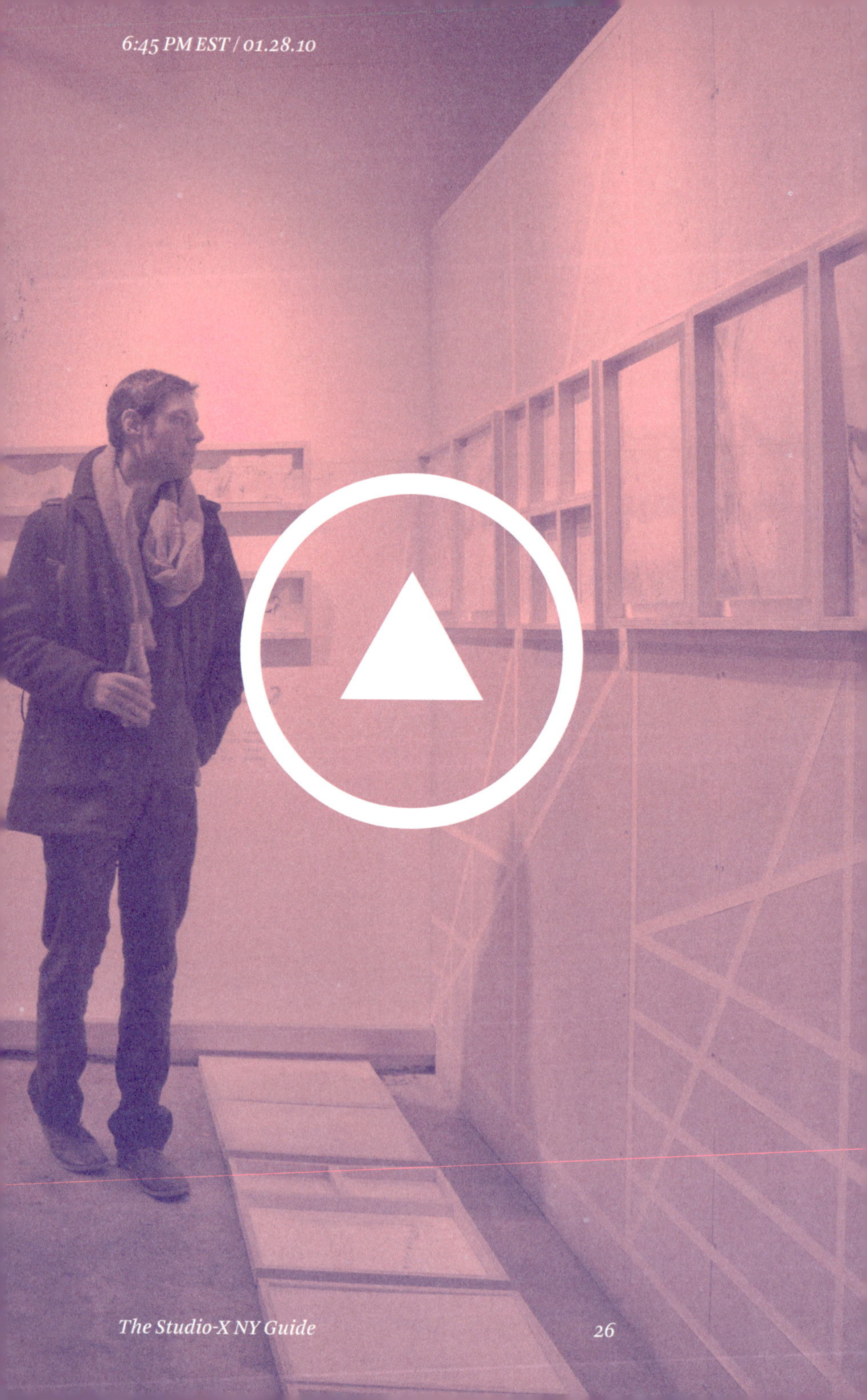

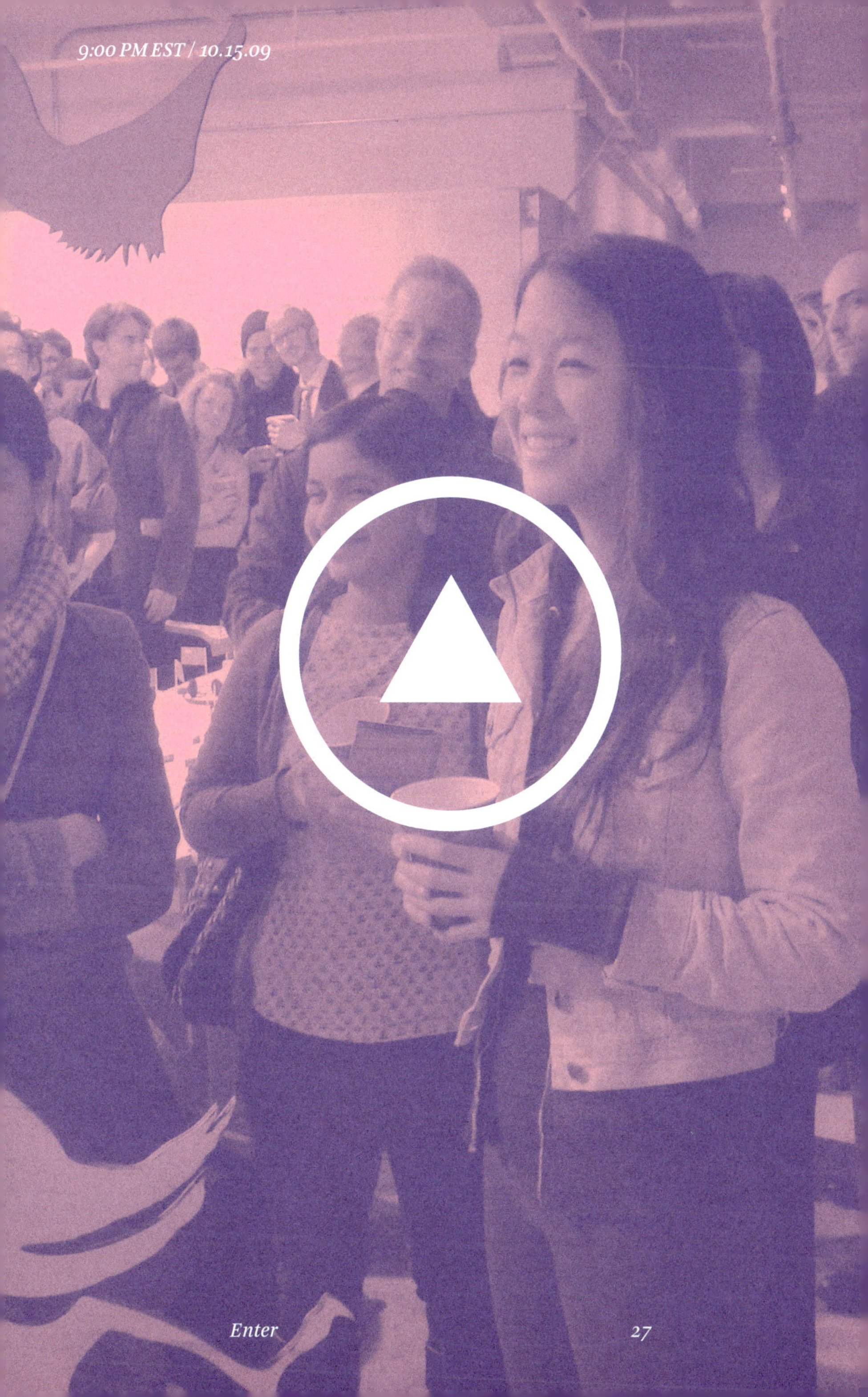

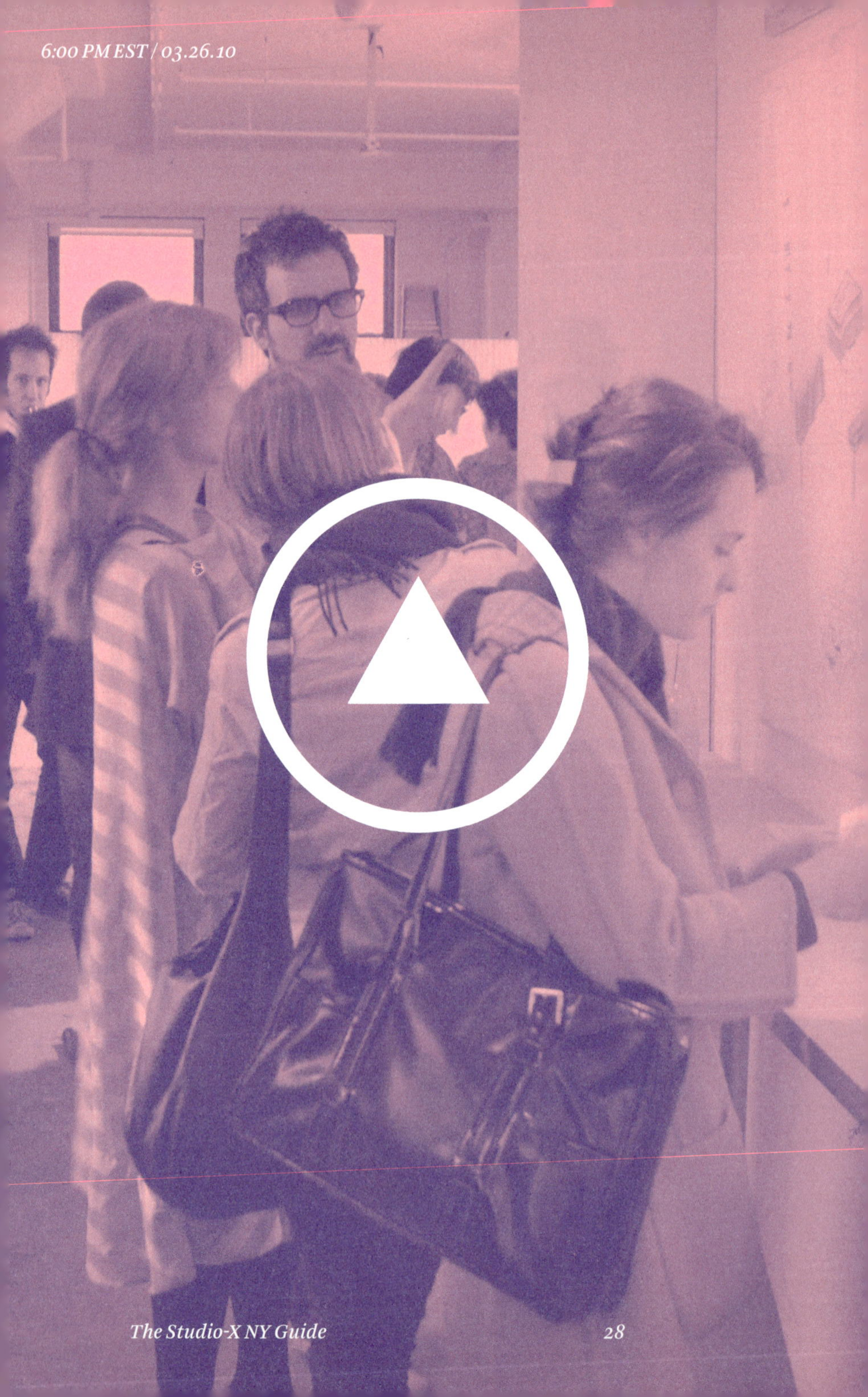

Classify.
Over the course of two years at
Studio-X New York, a range
of typologies emerged to classify
the events that took place. Many
are familiar—borrowed and
repurposed from the nightly news,
the tourism and entertainment
industries, or community
politics—and some are new. Each
typology takes on new meaning
according to the event it describes,
and each has its own icon.
Here are twenty-three examples,
with room to expand.

<u>**Artists Talk**</u>. Introduce practitioners from different disciplines. Ask them to present their work, and to discuss each other's work. The conversation may focus on method, materials, scale; it may highlight similarities or differences. What cross-disciplinary common ground emerges between a designer and a dancer? A sculptor, a performance artist, a painter? Which artists' works is already in conversation? The studio can enable this face-to-face conversation to take place.

See also: Dispatch

<u>**Book Launch**</u>. Launch a book by inviting all parties involved in its creation to share their thoughts: author, editor, designer, photographer, marketer—the more, the better. Present a cross-section of knowledge about how the book was assembled. Sell copies, and be inclusive by suggesting that all involved sign the book, not just the author. Do this for a new book, or re-launch an older book on its anniversary, or around a theme with contemporary relevance. A book launch should create conversation first, commerce second.

See also: Panel, Publishing Practice

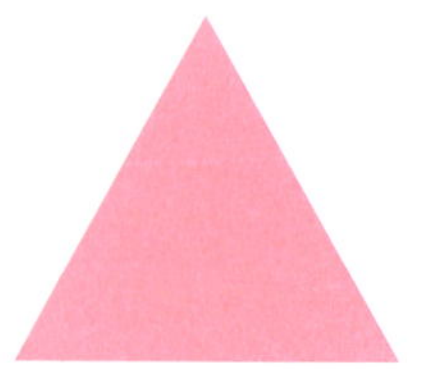

<u>Charrette</u>. Some problems need to be worked out with paper, pens, markers and conversation over the course of an afternoon. Identify different aspects of a problem, and assign a group to tackle each. Give each group specific conditions to address, but not orders. At the end of the day, pin up all ideas and engage in a group presentation and critique. Use this information as a platform for the next stage of your inquiry into a given topic.

See also: Crit, Publishing Practice, Workshop

<u>Competition</u>. Devise a competition. It can be a one-off or it can occur on a regular basis. Ground your brief in current events, and maintain a low entry fee. Define submission specifications, a deadline, a fair and transparent means of judging the winner, and a prize. Once entries are submitted, create an online exhibition. Later, create a real exhibition out of the entries: pin everything up on the walls, and invite experts from appropriate fields to discuss. Then, announce the winner.

See also: Exhibition, Rapid Response

<u>Crit</u>. Why should events only present end results? Celebrate incompleteness. Highlight process. Make a crit the main event. Bring a diverse group together to offer constructive criticism on works-in-progress. Down the road, showcase the work and the critique again, and highlight the directions it has since taken. Then, open that up for critique.

See also: Charrette, Free Speech Zone, Simulcast, Workshop

Dispatch. The international circuit of art and architecture biennales, triennales and fairs: each is place-specific, but what could be more democratic than bringing souvenirs of these spectacles home for a new audience? After the events close, invite the curators or participating artists, architects or filmmakers to discuss their experience. Don't aim to be comprehensive. Instead, tell a new narrative: if an architecture biennale involved film in a secondary or tertiary way, then foreground film. *Dispatches* highlight ephemera and personal reflection, and place them within a new context.
See also: Artists Talk, Panel

Exhibition. A studio can have multiple functions. In addition to walls, try using doors, ceilings, pipes, windows or the floor as surfaces to display work. Once work is hung, consider these exhibitions works in-progress: conversation pieces for further inquiry and public programming. An exhibition can be the basis for many things: a series of works on display, a panel to discuss it, a party to launch it, a workshop to develop it further, a growing archive and publication to house it, and a body of work to share.
See also: Competition, Panel, Publishing Practice, Workshop

<u>Free Speech Zone</u>. Some people don't like cameras. Why diminish their enthusiasm or desire to speak freely by recording every single event? Eliminate the fear of reproach. Turn off the camera. Be off-the-record. Close the door. Open the floor to conversation.

See also: Crit, Group Therapy, Panel, Rapid Response, Reading Group, Simulcast, Workshop

<u>Group Therapy</u>. Economic, social, and spatial processes shape our everyday lives. But so do emotional ones— both individually and collectively. How do we move through the world, together? Create a safe environment for people to process their feelings about the spaces around them.

See also: Free Speech Zone, Panel, Rapid Response, Workshop

<u>Neighborhood Watch</u>. We are all local experts. Look to your immediate surroundings. Invite those around you to share their lives, work, and interests. Create new networks within your building, street, block or neighborhood (and in your personal life) by showcasing the work of those who live and work there. What are they reading, writing, designing, thinking, eating? What is important to them, and why?

See also: Look Local, Mini Series, Publishing Practice, Tour

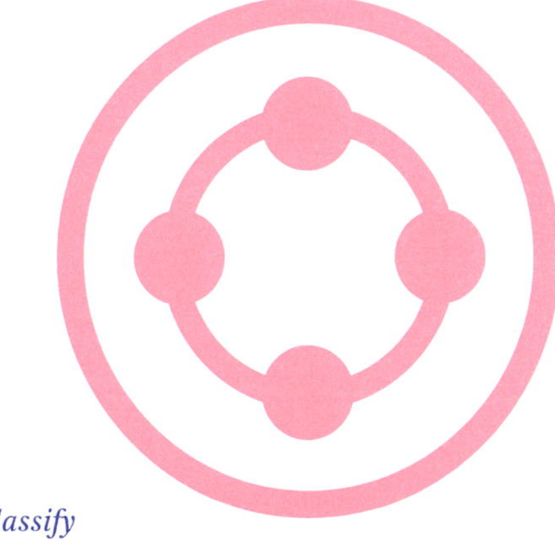

Look Local. As a concept, *Look Local* untethers *Neighborhood Watch*, allowing it to hone in on other microgeographies. These sites may be located in an adjacent neighborhood, in another city or on an opposite coast. *Look Local* zooms in, questioning what forces, activities, policies and procedures give form and identity to locales.

See also: Neighborhood Watch, Tour, Off-Site

Mini Series. Choose a broad topic. Investigate it at regular intervals over time, from different angles and with different parties. This will create a rhythm over the course of six months or one year. Pick an issue of interest to a broad swath of people, or one that affects your local environment. At the end of the series, consolidate the ideas generated into a publication or policy recommendation.

See also: Neighborhood Watch, Panel, Pile-On, Publishing Practice, Town Hall Meeting

Off-Site. Collaborate with another institution: cross-promote, bring your perspective and audience to its work, and hold an event at its venue. Foster interdisciplinary conversations among people and institutions. Build new alliances and audiences in this anti-*Dispatch*. See how your studio feels in someone else's space.

See also: Dispatch, Look Local, Tour

<u>Panel</u>. A panel is both an entity unto itself and a building block. It can be the main event, or it can stack with other panels into an evenly paced *Mini Series* or a frenetic *Pile-On*. *Panels* can address any topic; they can be quick and dirty or thoughtful and researched. The permutations are endless, requiring only a plurality of opinion and perspective. Without that, it wouldn't be a panel.

See also: Book Launch, Free Speech Zone, Group Therapy, Mini Series, Pile-On, Publishing Practice, Rapid Response, Town Hall Meeting

<u>Performance</u>. Why can't a studio be a stage? Give the main area over to an individual or a group. Be open to all art forms. Ask performers to create a new work, allowing them as much advance rehearsal time in the studio as they need. Require only experimentation and site-specificity—but encourage participatory projects.

<u>Pile-On</u>. Why pace yourself? Accelerate and condense a *Mini Series* into one congested, high-density afternoon. Transform your studio into a space momentarily and wholly dedicated to one given topic. Challenge yourself to have as many disciplines as possible represented. Sell relevant books and periodicals. Time sessions to be back-to-back. Exhaust the audience. Overload them with information. Pack the house.

See also: Panel, Mini Series

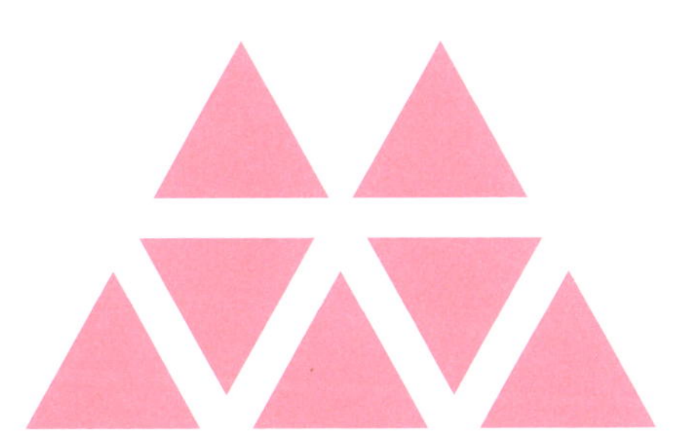

<u>**Publishing Practice**</u>. **Transform your studio into a reading room. Invite your neighbors and studiomates to display anything they've recently published—from books to articles to a take-out menu from the restaurant downstairs —for a one-day exhibition. Or, hang publications on the wall with Velcro for an extended display. Transcribe, edit and excerpt events into annual or sporadic publications. Invite people to use the studio as a space to workshop their ideas, scripts or unfinished books in various formats— print-on-demand, audiobook, PDF.**
See also: Book Launch, Charrette, Exhibition, Mini Series, Neighborhood Watch, Panel, Workshop

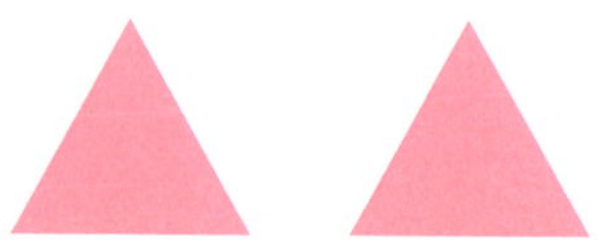

<u>**Rapid Response**</u>. **Reserve at least one evening per month to respond to current events, announcing in advance only the date of the response, not what event or topic partici- pants will be responding to. When something major occurs, consider various means of response that might be helpful to others, either analytically or through the use of the studio space itself. Does a spike in unemployment rates, for example, necessitate an ad hoc job fair at your studio? Do contested policy decisions require information sessions and interventions? Do natural disasters and relief efforts require idea-generating competitions?** *Rapid Response* **is an ongoing conversation with the events that shape our lives.**
See also: Competition, Exhibition, Free Speech Zone, Group Therapy, Panel, Town Hall Meeting, Workshop

<u>**Reading Group**</u>. **What book would you like to explore with others? Ask five to six people to commit to reading it with you. Find an organizing question, and create a syllabus. Assign supplemental materials such as newspaper clippings. Meet on a weekly basis. Screen relevant film clips to spur conversation. In the final session, invite the author to the studio for a capstone conversation with the group. Afterward, ask all members to write a reaction paragraph about reading the book together, the experience of meeting the author, and the book itself. Share these reaction paragraphs among the group, and send them to the author.**

See also: Free Speech Zone

<u>**Simulcast**</u>. **Why should events be constrained to one space? To one audience, culture or language? Simple communication technologies enable the participation of speakers in other countries and time zones. Find a like-minded partner organization in another city or country. Ask them to invite speakers and an audience. Create informal, cross-cultural dialogue on a variety of topics by encouraging not only panelists, but also audiences, to interact.**

See also: Crit, Exhibition, Free Speech Zone, Workshop

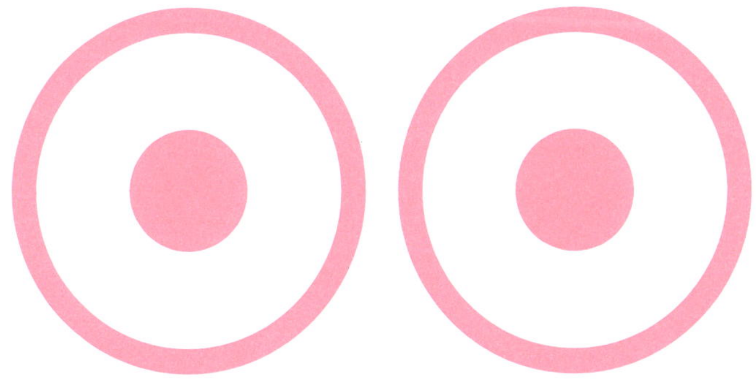

<u>Tour</u>. Leave the studio and keep moving. A tour can be organized in many ways. It can be themed or specific to a time of day or time of year. It can be data-heavy, historic, sonic, chronological or aided by technology. It may or may not have a destination. Organize a tour for any time of day or night. Engage the city in a new way, through the eyes of a guide.
See also: Look Local, Neighborhood Watch

<u>Town Hall Meeting</u>. Why can't a studio affect policy decisions, or at the very least, be a place where these decisions are discussed by those affected? Find a topic that impacts many different stakeholders—neighborhood groups, city agencies, the business community—and invite representatives from each to share their perspectives. Acknowledge that these issues are often sensitive by creating a neutral environment and by inviting an informed moderator. Advertise to the public, and encourage participation. Give everyone equal time and equal respect.
See also: Charrette, Look Local, Group Therapy, Mini Series, Panel, Rapid Response, Workshop

<u>Workshop</u>. Why should an event have a neat conclusion? The workshop process resists drawing conclusions, posing more questions than it answers. Workshops can be visual or verbal, and can build new coalitions by bringing different parties together to discuss any given topic.
See also: Charrette, Crit, Exhibitions, Free Speech Zone, Group Therapy, Publishing Practice, Rapid Response, Simulcast, Town Hall Meeting

Transform.
Studio-X New York is not just
an event space—it is also a working
studio. Here is how to transform
a workspace into an event space
in thirty minutes.

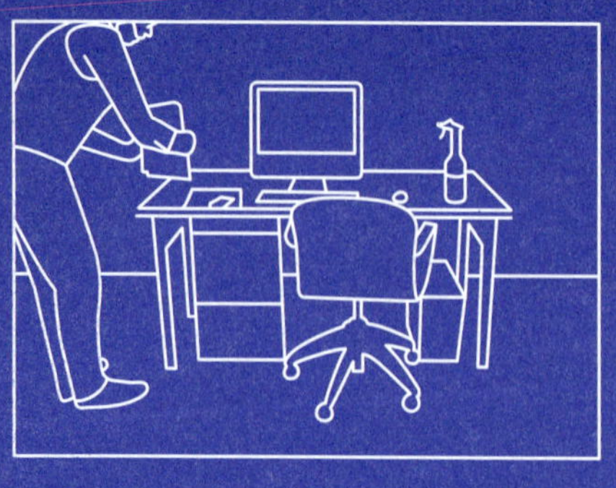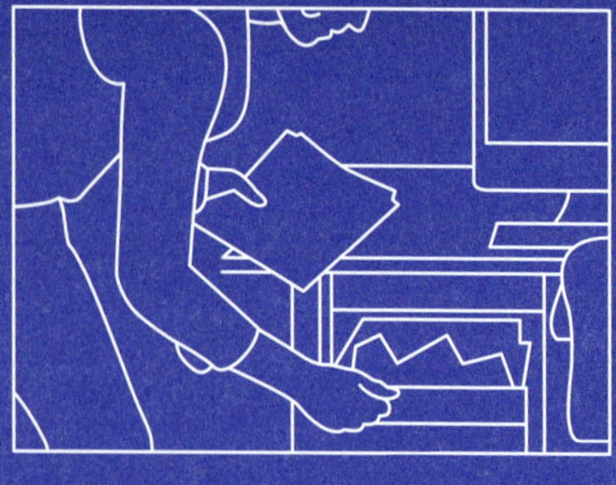

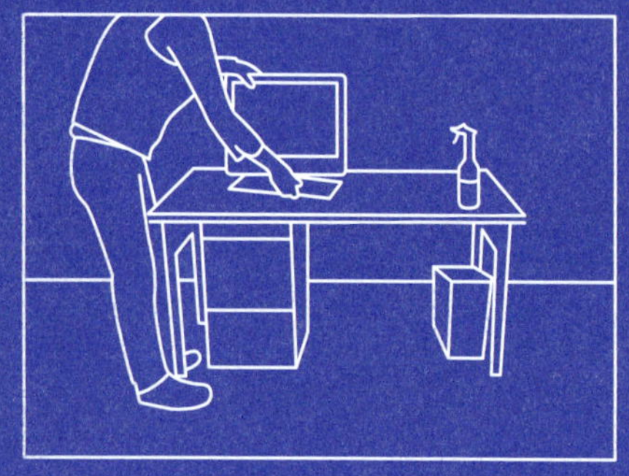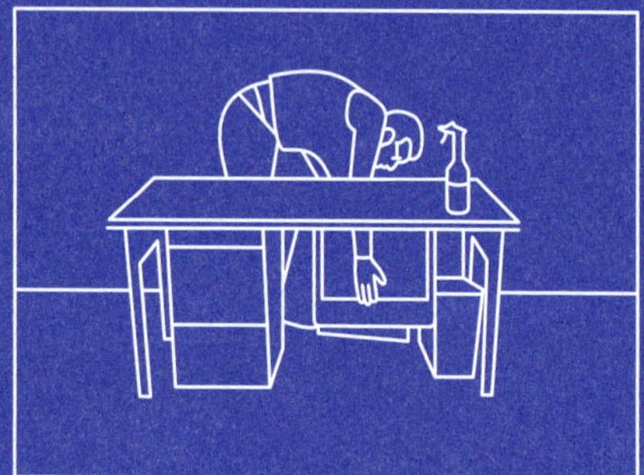

Store

Begin by removing papers from working surfaces and storing them in locked file cabinets. Everyone who works in the space should be assigned a file cabinet, and each file cabinet should have wheels. This will allow for greater flexibility when arranging the room. File cabinets can be rolled anywhere in the space to accommodate different arrangements and uses. They can also serve double-duty, offering storage and surface areas for speakers to rest their notes and drinks while presenting. Don't be afraid to incorporate everyday infrastructure into public programming.

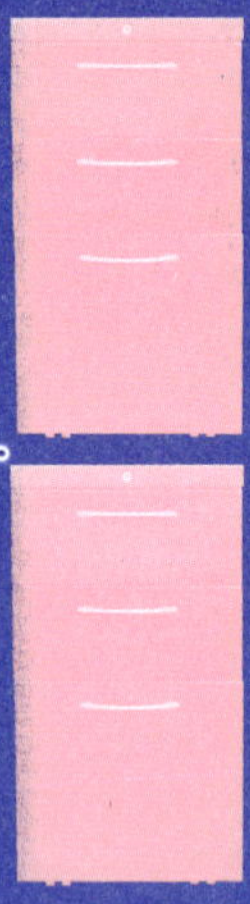

item: file cabinets
brand: HON
style: Embark™ Standard Height Pedestals
collection: Embark™ Mobile Pedestals
size: 1'8" L x 1'3" W x 2'1' H
color: black
cost/unit: $352
qty: 14

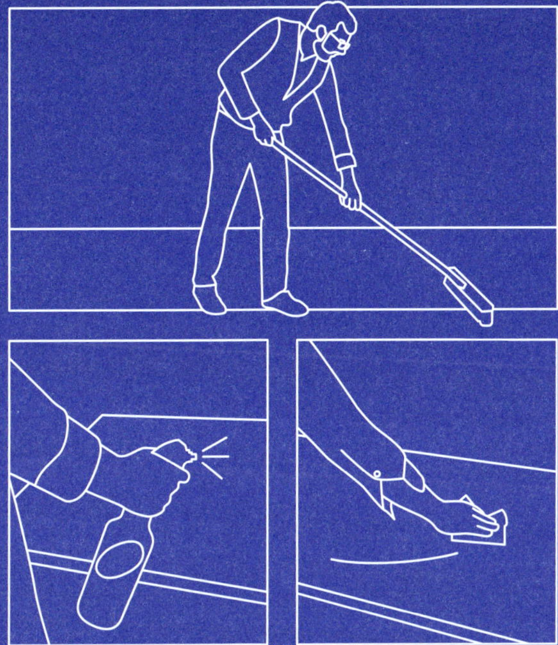

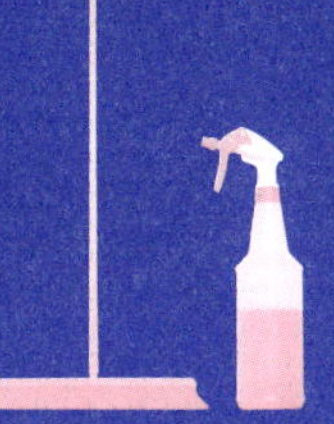

item: broom
qty: 1

item: cleaning fluid
brand: Simple Green
Concentrated All-
Purpose Cleaner/
Degreaser/Deodorizer
size: 1 gal.
cost: $19.99
qty: 1

item: spray bottle
brand: Unisan
size: 24 oz.
cost: $3.49
qty: 1

item: closet
est. materials cost:
$579.99
(breakdown below)
qty: 1

item: wood
type: baltic birch
plywood
size: .75" L x 4' W x 8' H
cost/sheet: $55
qty: 11 sheets

item: hinges
cost: free; salvaged
materials
qty: 4

item: screws
brand: McMaster
style: thread-forming
screws
length: various
cost: $5–$10/box of 100
qty: 559

Clean

Once all surfaces are cleared, clean
them! Purchase a large bottle of concentrated cleaner. As needed, transfer it
into a spray bottle and dilute for light
cleaning. This ritual of spraying and
wiping down all surfaces will not only
freshen things up and signal a transi-
tion for the person in charge, it will also
send a message to all in the space that
an event is right around the corner.

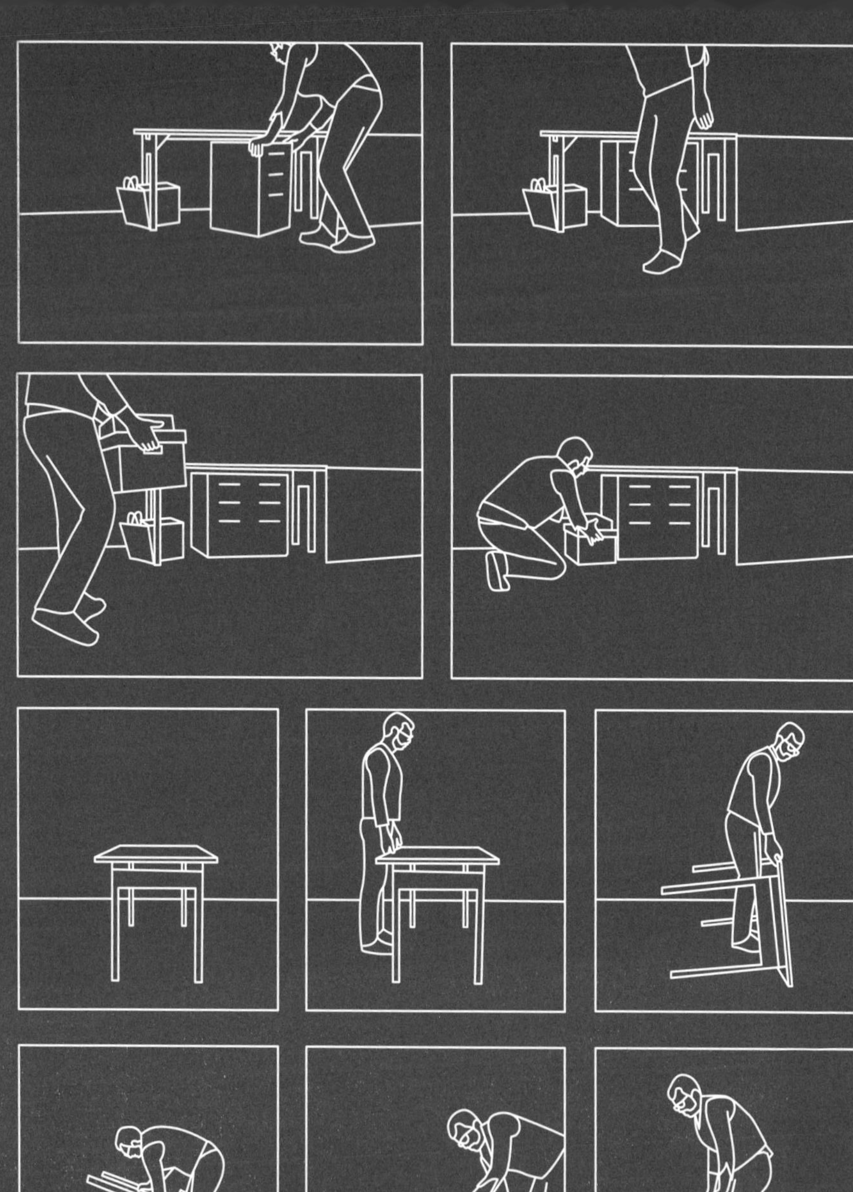

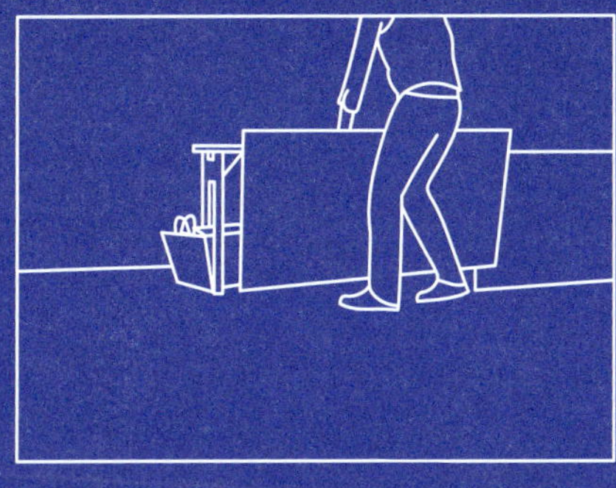

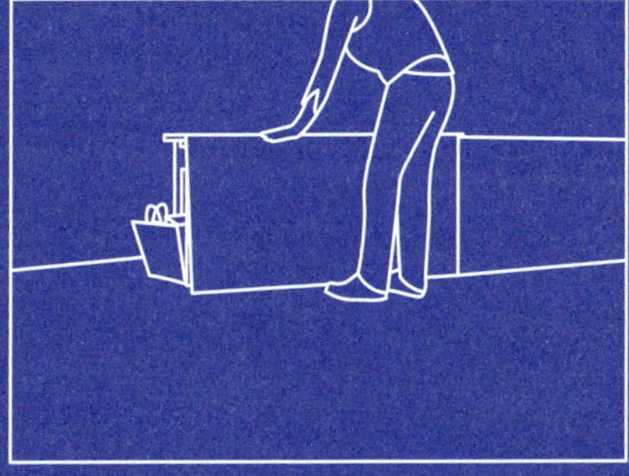

Hide

Sturdy folding tables are not only great for working, they're also essential to a successful event. At night, tables take on various roles at Studio-X New York: bar, buffet table, infoshop, bookstore, and more. But when you are setting up for an event, their most crucial function may be to hide things. Once you've cleared and cleaned your surfaces, place everything (including your computer) under your table. Fold another table flat and rest it against the first. This will secure your items while providing spillover seating for those who come late to an event. Ask everyone working in the space to do this.

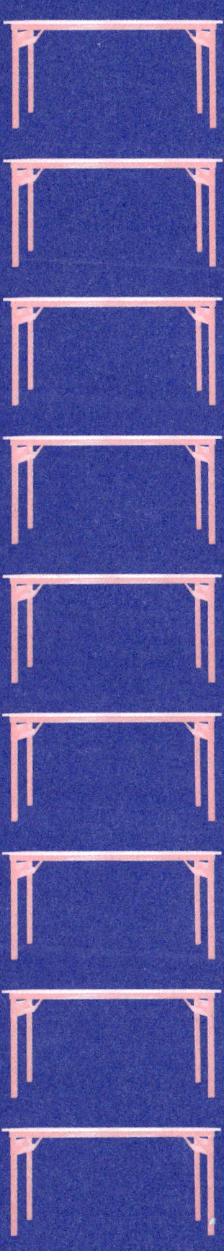

item: tables
brand: Maywood
style: Original Series:
Rectangles; black,
high pressure laminate
size: 30" L x 60" W
cost/table: $150
qty: 28

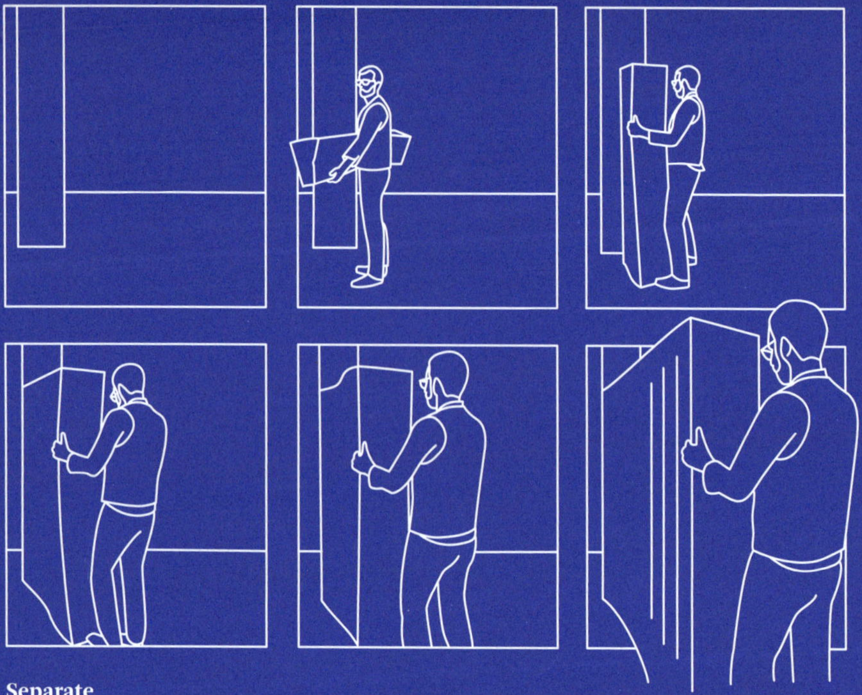

Separate

In a busy design studio, the workday doesn't always end at 6pm. For those who need to continue working, a mobile paper wall allows for quick separation of the room. That way, people who don't wish to stow and hide their items may pack up and move behind the wall, and work quietly and privately behind this temporary division. The wall creates an important separation between work and event space, and allows both to operate simultaneously.

item: paper wall
brand: molo design, ltd.
style: softwall White Textile
size: 12" W x H' L
cost: $2,000
cost with educational discount: $1,000
qty: 1

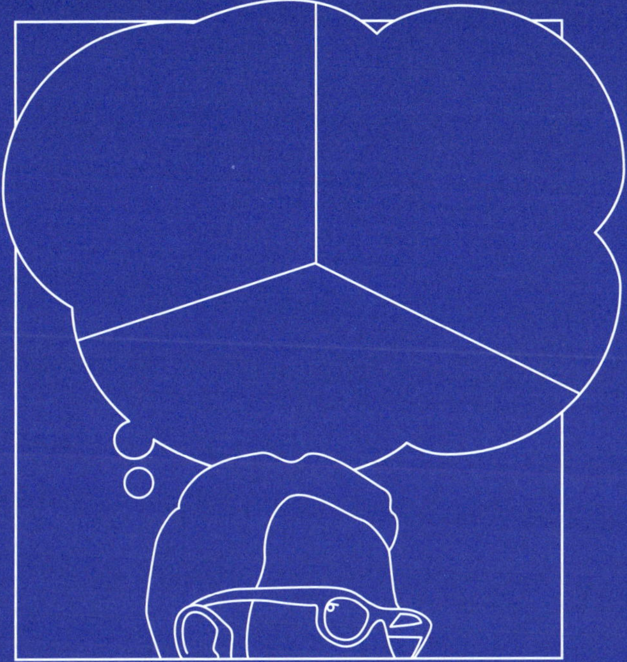

Imagine

Now that you've prepared for the event, figure out how to use the space. Which direction will the audience face? Which direction will the speakers face? Will they face each other, or will another configuration arise? Will speakers and audience be seated or standing? Will the event be quiet or loud, still or dynamic? Regardless of how the space is arranged, unfinished and untreated surfaces allow for maximum flexibility. White walls that may minutes before have been used as a pin-up space are now used as a screen on which to project images, or as the backdrop and set for a performance. Every mutable surface takes on a temporarily assigned meaning.

item: wall paint
brand: Benjamin Moore & Co
style: Regal Flat Finish
color: Decorator White, N215-04
size: 1 qt.
cost: $14.10/qt.

item: floor
material: concrete
cost/biweekly cleaning: $150
qty: 1

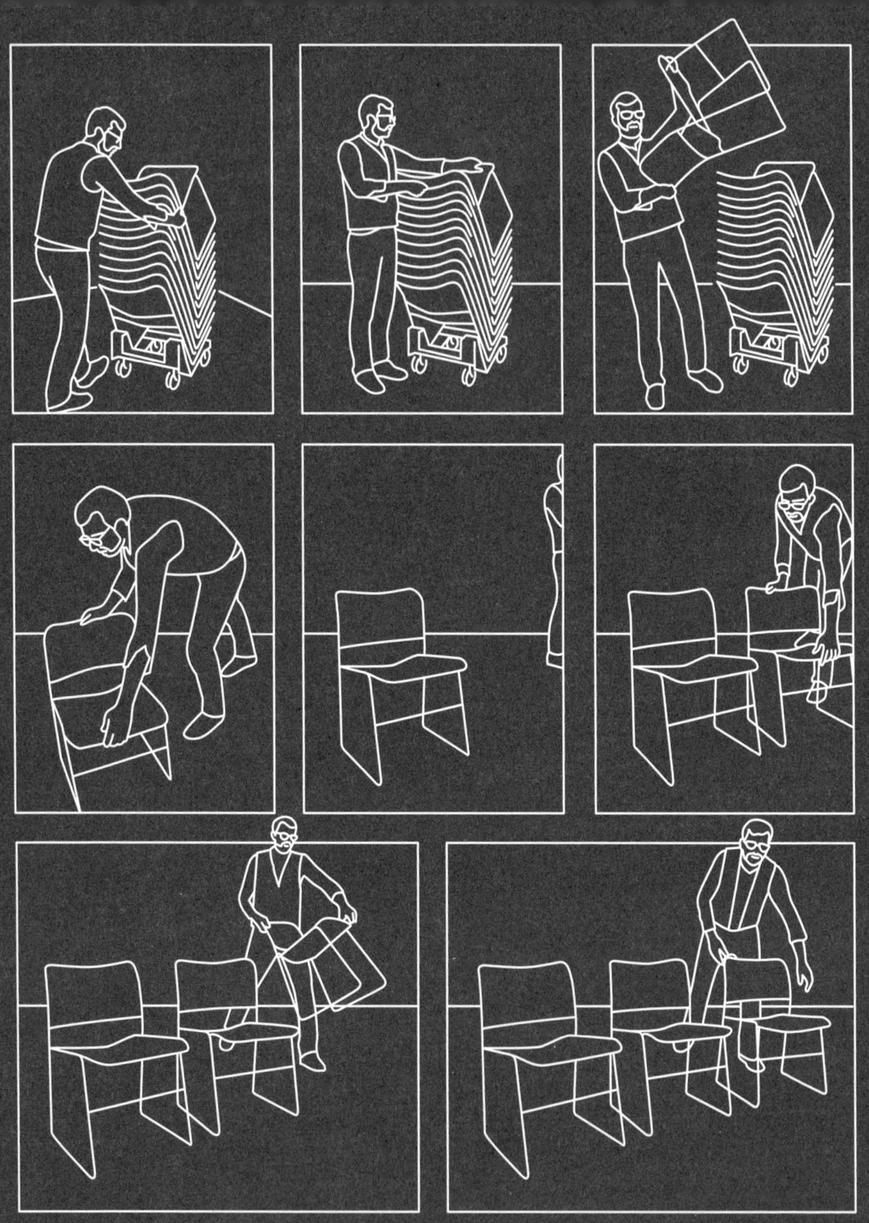

<u>**Arrange**</u>
Stacking chairs are easy to store, and a
dolly allows them to be easily hidden
or moved around the space. Chairs can
then quickly be unstacked and arranged
in whatever formation seems right
for a particular event: a circle, a square,
rectilinear or radiating rows, or something entirely new. Set a few chairs aside
for latecomers. Experiment with room
arrangements, and see how many con-
figurations you can discover!

<u>**Arrange Some More**</u>
That takes care of the audience. Now,
what about the speakers? Will they
stand or sit? Will they sit behind a table
or next to a file cabinet? Many factors
must be considered: for example, the
length and format of the program,
and whether or not it will employ sound
or images. A workshop will require different speaker-audience dynamics than
a panel discussion or a reading group,
and this is altogether different from
what might be required by a charrette.

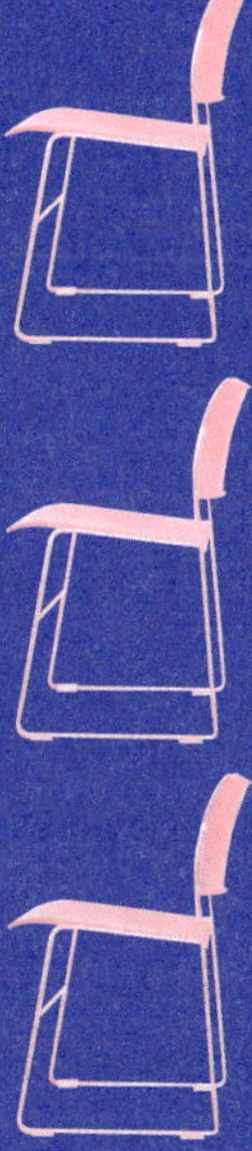

item: chair
designer: David
Rowland
style: 40/4 chairs
size: 19.75" x 30"
seat height: 17.5"
seat depth: 18"
cost/chair: $159
qty: 48

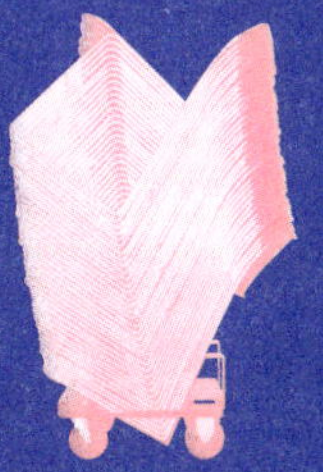

item: 40/4 Chair Dolly
designer: David
Rowland
size: 20" x 22"
cost: $149
qty: 1

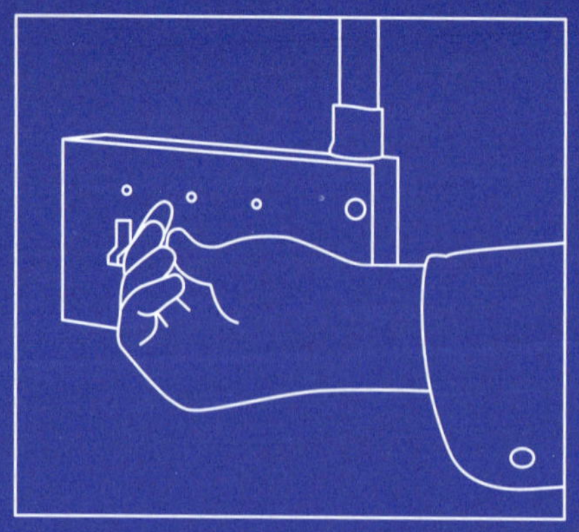

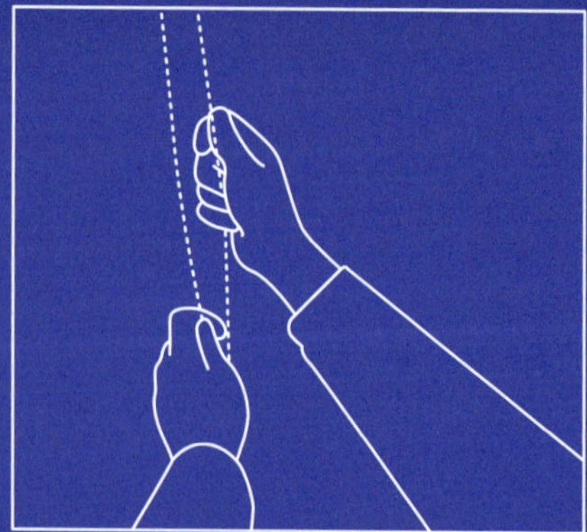

Adjust

What a difference lighting makes! If
overhead lighting connotes "work,"
then a simple flip of the switch to track
lighting signifies "event," and does
so while still providing enough light for
people to comfortably work behind a
division (see *Separate*). During the long
days of summer, you might not need
lights at all, but a simple set of shades
(and for daytime events that use images,
blackout curtains to darken the space).
With whatever seasonally appropriate
option you choose, a dimmed, cool
space is welcoming. It makes people
feel that they are not walking into
an office, but a place where something
is about to happen.

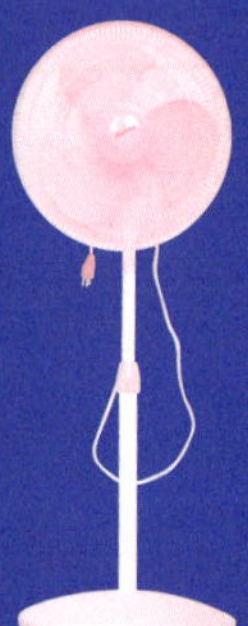

item: fan
brand: Air King
style: Oscillating Electric
Pedestal Fan
type: 16-1P
model: 9126M 4C945M
size: 16"
cost: $69.95
qty: 2

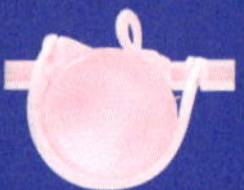

item: track lighting
heads
brand: WAC
style: LTK-764
size: 5" W x 5.187" H
cost/head: $19.71
qty: 18

item: track lighting
bulbs
brand: GE
style: Reveal 75 Watt
Indoor Flood Beam
lumens: 950
cost/bulb: $9
qty: 18

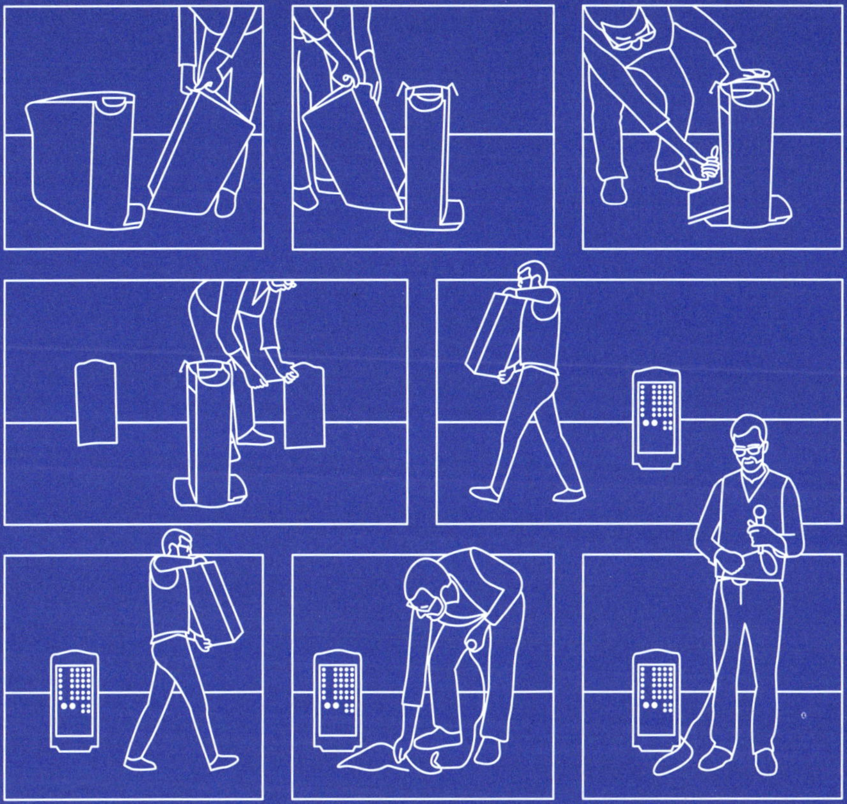

Communicate

Audio-visual presentations: they can be the most seamless or the most stressful aspect of an event. Allow ample preparation time. If an event requires images, set up the projector, and decide in which direction to project. Will the projector hang from the ceiling, or rest on a table? A portable PA system is also useful, allowing for easy storage. Unpack it and arrange speakers according to the layout of the room. Have microphones on hand. Begin playing music fifteen minutes in advance of the event's start time, knowing that some people will arrive early.

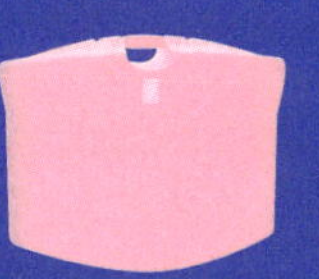

item: public address system
brand: Fender
style: M123
size: 12" L x 15" W x 128" H
cost: $878
qty: 1

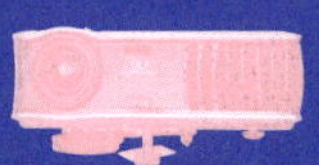

item: projector
brand: Dell
style: 2400MP DLP Front Projector
lumens: 3000
cost: discontinued
qty: 1

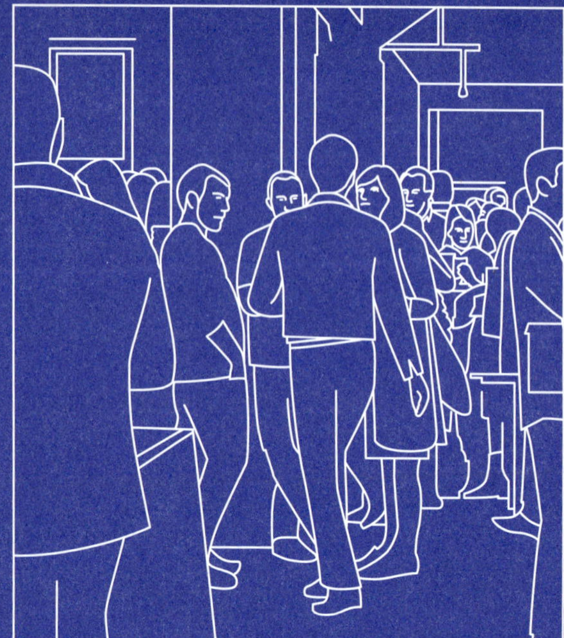

Welcome

Greet guests as they enter. Offer them drinks. Make sure that the bathroom key is handy and well marked. Once everyone is settled, introduce speakers and hand over the floor. Watch and listen. When you feel that the event has gone on long enough, send a not-so-subtle signal that it's time to wrap things up: move to the back of the space, then loudly pop open soft drinks and bottles of wine. Applaud. Turn music back on. Mingle.

item: shelf
est. materials cost: $264
(breakdown below)
qty: 1

item: wood
type: .75" baltic birch plywood
size: .75" L x 4' W x 8' H
cost/sheet: $55
qty: 2 sheets

item: flanges
brand: Speed Rail
style: No. 45SBC Square Base
cost/flange: $9
qty: 7

item: pipes
brand: unspecified
total size: 27"
cost/foot: $3

item: screws
brand: McMaster
style: thread-forming screws
length: various
cost/box of 100: $5–$10
qty: 113

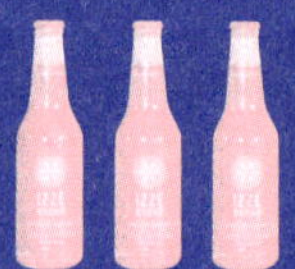

item: beverages
brand: IZZE Sparkling Juice
flavors: Birch, Blackberry, Mandarin, Lime, Pink Grapefruit, Pomegranate
size: 1 bottle (355 ml)
calories: 120
cost: in-kind donation
qty: 24/case

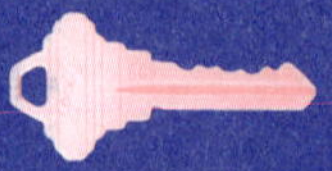

item: bathroom keys
brand: Kaba-Ilco SC9
cost/key at Greenwich Locksmiths: $2.75
qty: 2

Stage.
How many ways can a space be
arranged? Does form follow
content, or vice-versa? Here are
thirty-seven examples.

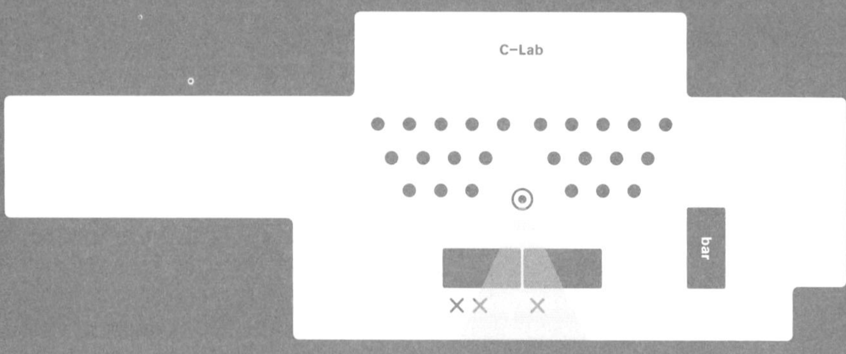

Rapid Response: Collapse!
10.28.08
C—Lab
bar

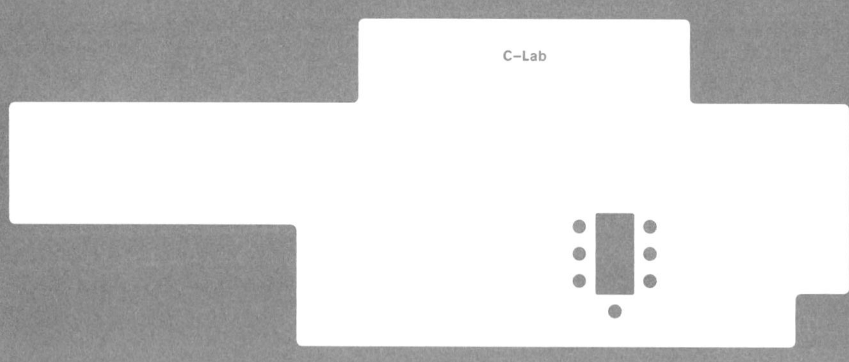

Reading Group: Flying Close to the Sun: My Life and Times as a Weatherman
11.10.08
C—Lab

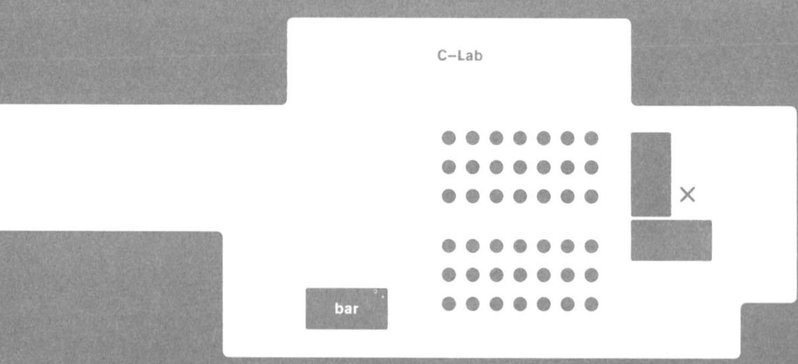

re:construction
11.18.08
C—Lab
bar

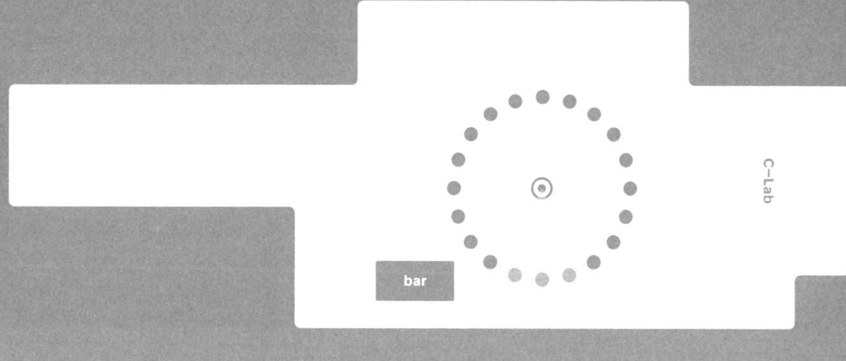

Rapid Response: I Need My Space
11.25.08
C-Lab
bar

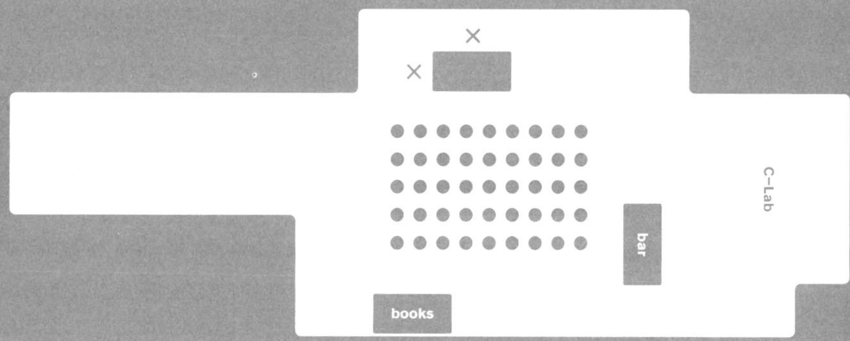

Night Haunts
12.04.08
C-Lab
bar
books

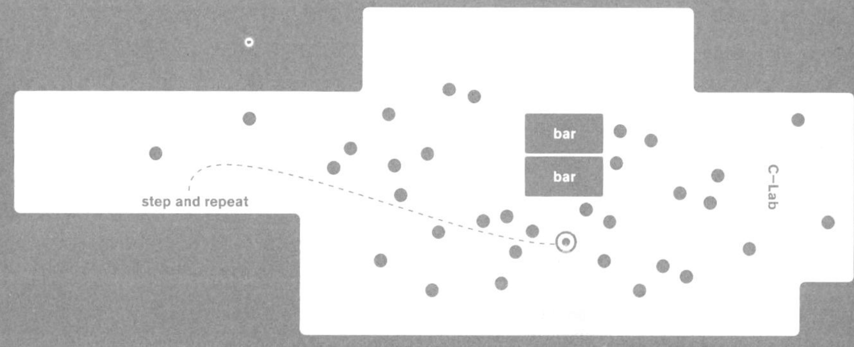

Volume 17: Content Management Launch Party
12.09.08
bar
bar
C-Lab
step and repeat

A Few Zines: Dispatches from the Edge of Architectural Production Opening
01.08.09

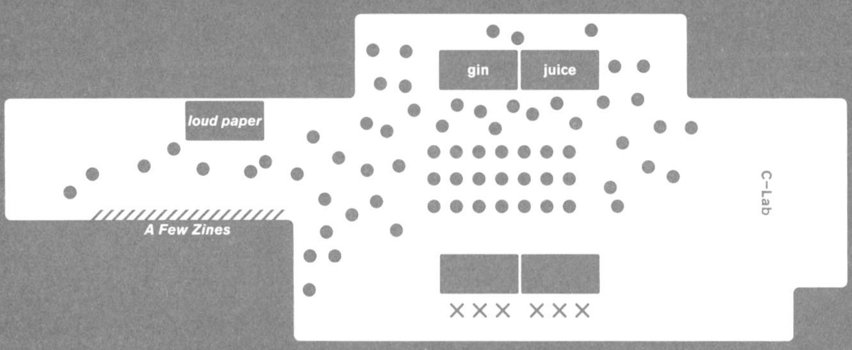

A Few Zines: Dispatches from the Edge of Architectural Production
01.08–02.28.09

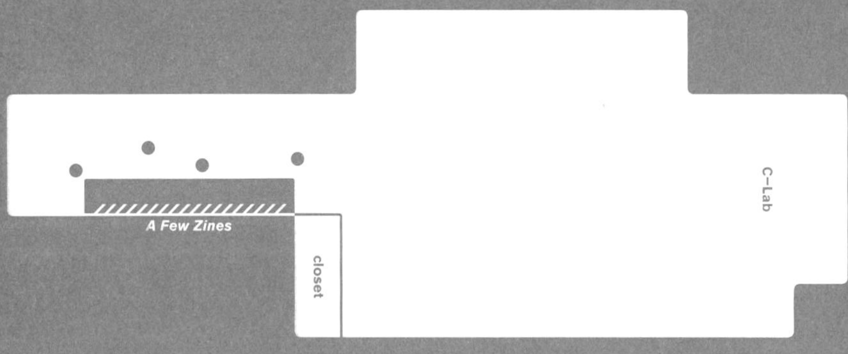

J. MAYER H.
01.30.09

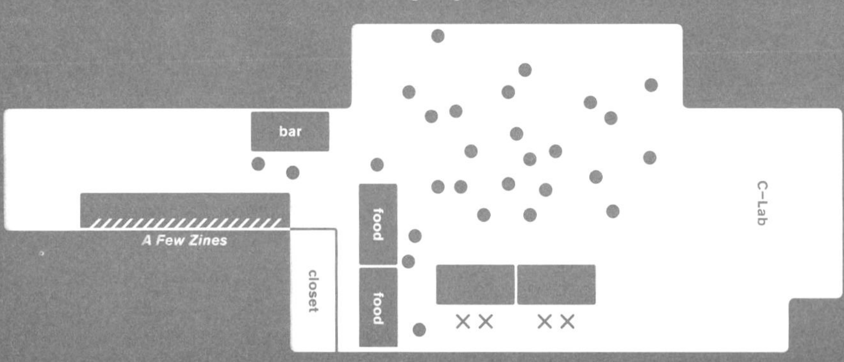

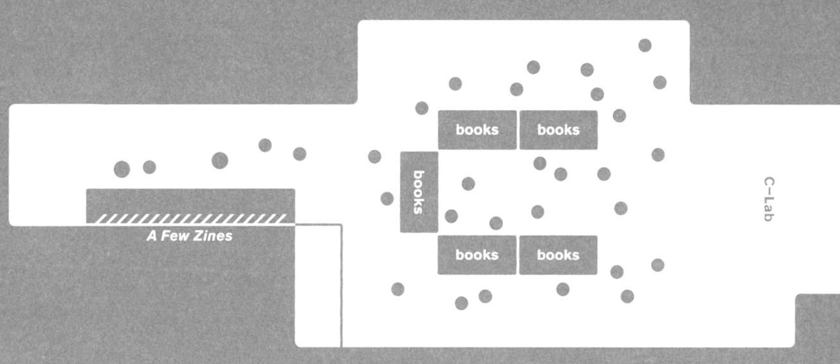

Neighborhood Watch: Writers' Block
02.19.09
books
books
books
books
books
A Few Zines
C–Lab

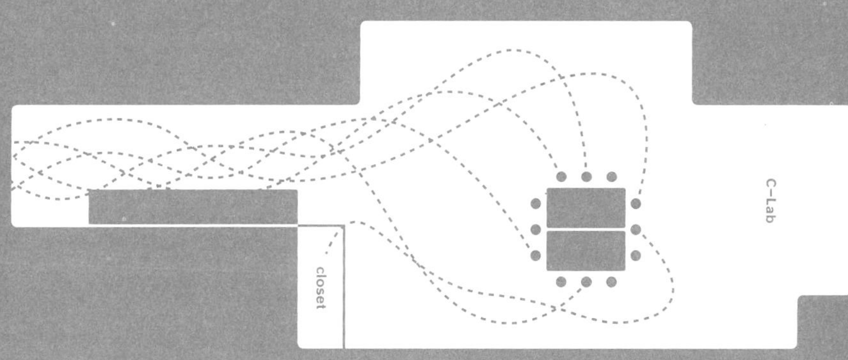

spurse: Deep Time + Rapid Time
03.03.09
closet
C–Lab

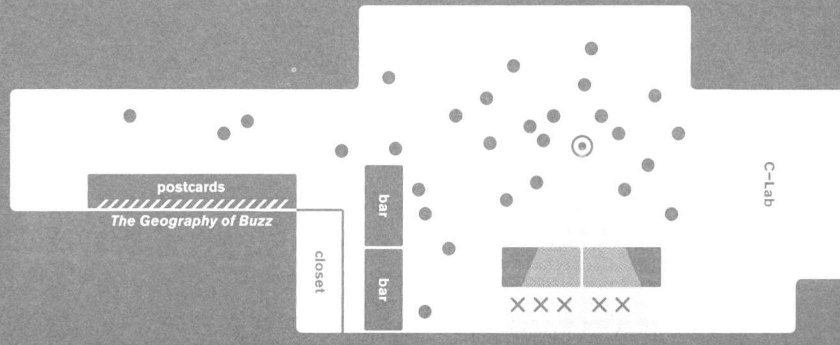

The Geography of Buzz Opening
04.07.09
postcards
The Geography of Buzz
bar
bar
closet
C–Lab

The Geography of Buzz
04.07–05.13.09

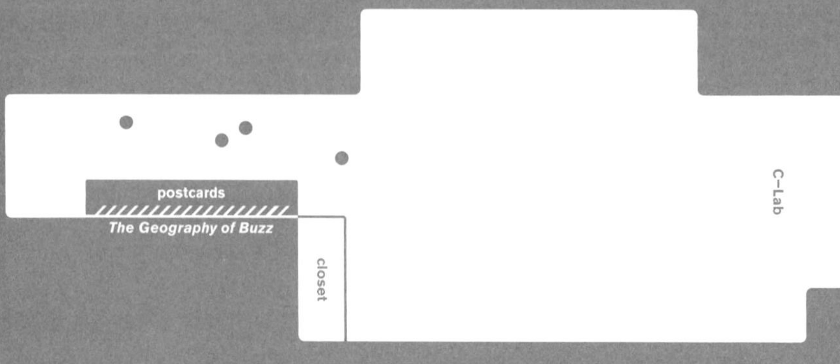

ISLANDS + GHETTOS
04.28.09

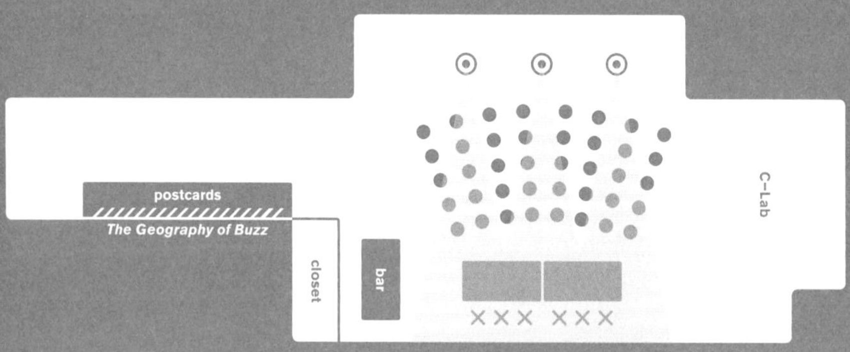

InDisposed: Talking Trash about Design Opening
05.14.09

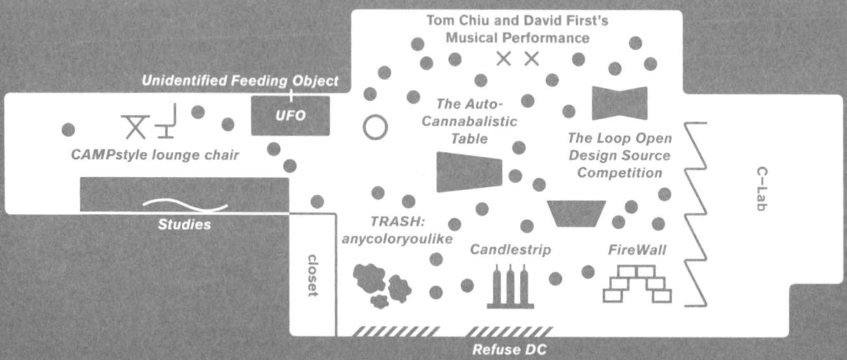

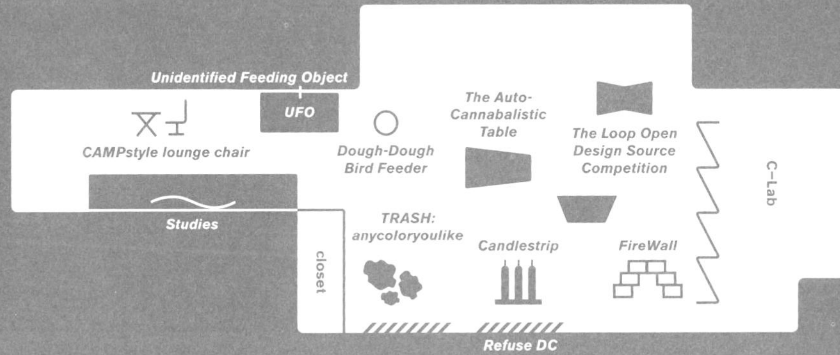

InDisposed: Talking Trash about Design
05.14–05.21.09
Unidentified Feeding Object
UFO
CAMPstyle lounge chair
Dough-Dough Bird Feeder
The Auto-Cannabalistic Table
The Loop Open Design Source Competition
C–Lab
Studies
TRASH: anycoloryoulike
Candlestrip
FireWall
closet
Refuse DC

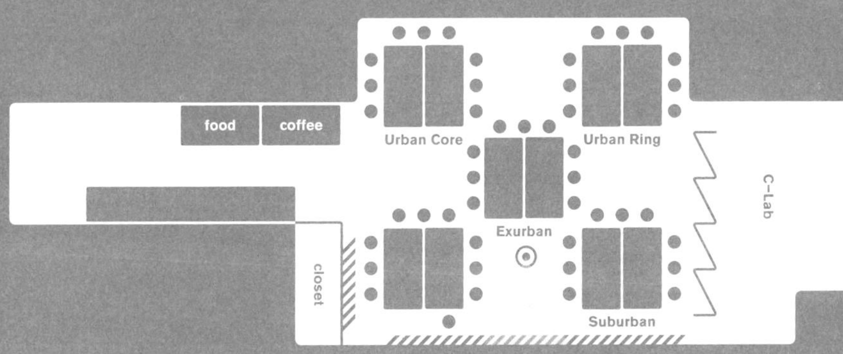

Public Housing: Policy and Design Workshop
06.12.09
food
coffee
Urban Core
Urban Ring
C–Lab
Exurban
closet
Suburban

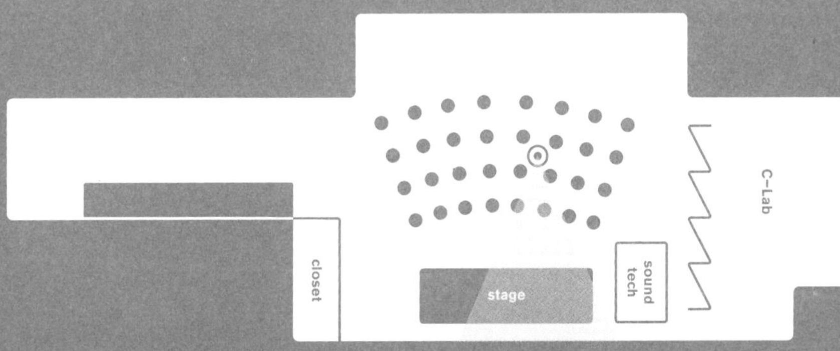

Eastern Western Meat Lessons
07.14.09
C–Lab
closet
stage
sound tech

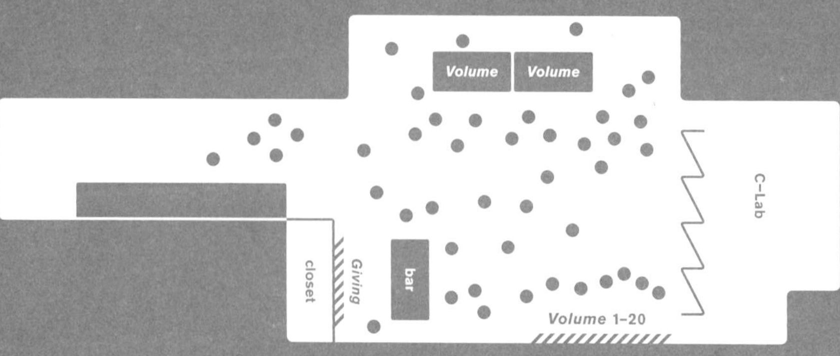

Volume 20: Storytelling Launch Party
09.19.09
Volume
Volume
closet
Giving
bar
Volume 1–20
C–Lab

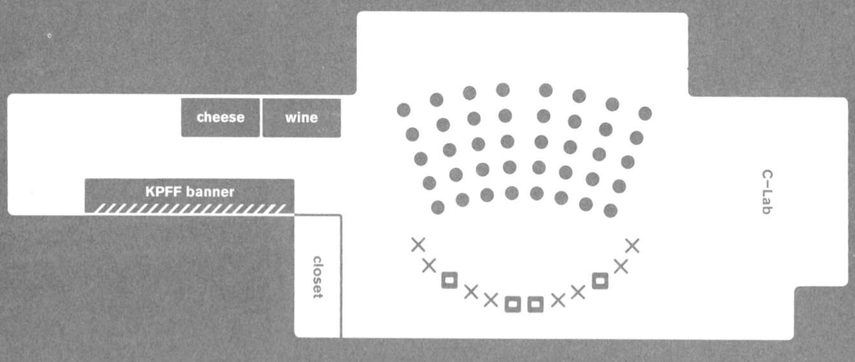

PRE-Office: Conversations with Architects
10.08.09
cheese
wine
KPFF banner
closet
C–Lab

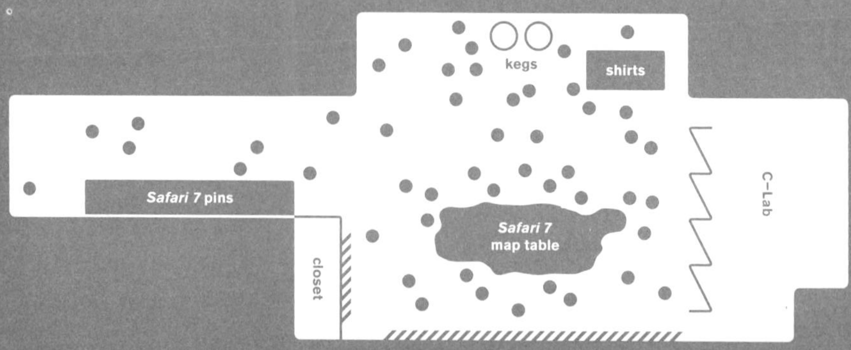

Safari 7 Reading Room Opening
10.15.09
kegs
shirts
Safari 7 pins
closet
Safari 7
map table
C–Lab

Safari 7 Reading Room
10.15–12.31.09

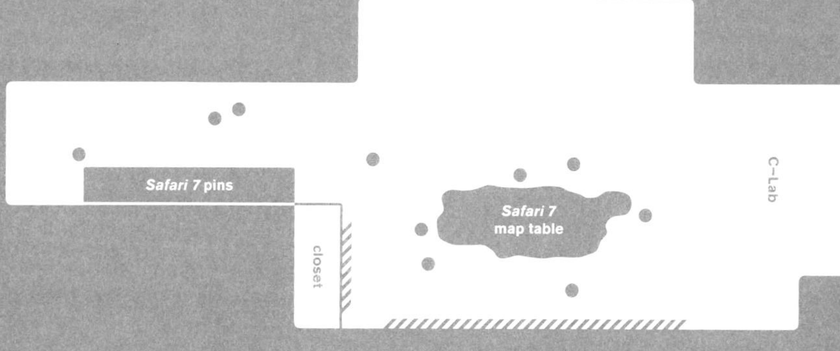

Landscapes of Quarantine Crit
10.24.09

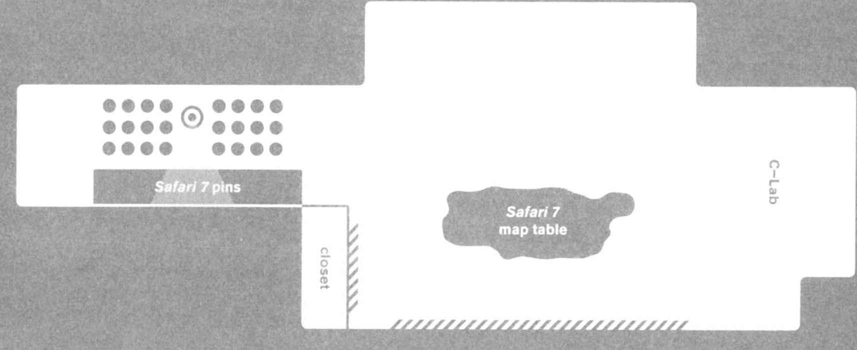

Dispatches from Villa Feuerlöscher
11.10.09

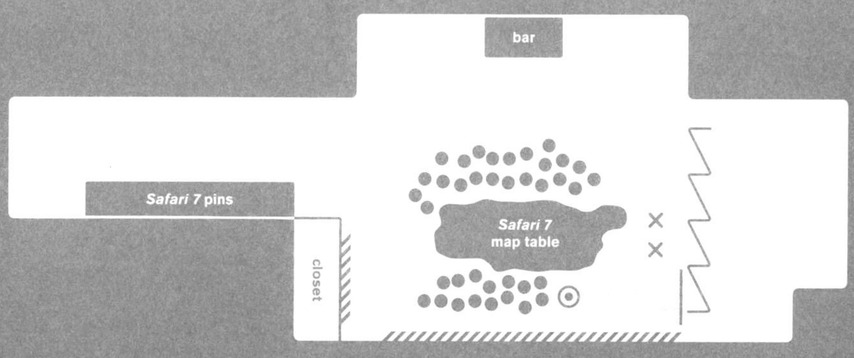

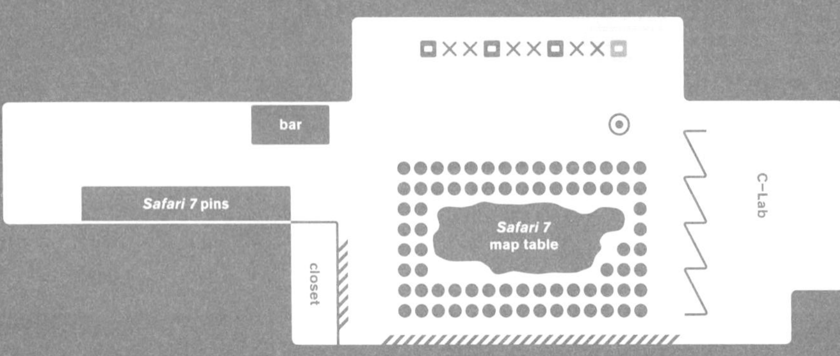

The Garment District
11.17.09
bar
Safari 7 pins
closet
Safari 7 map table
C-Lab

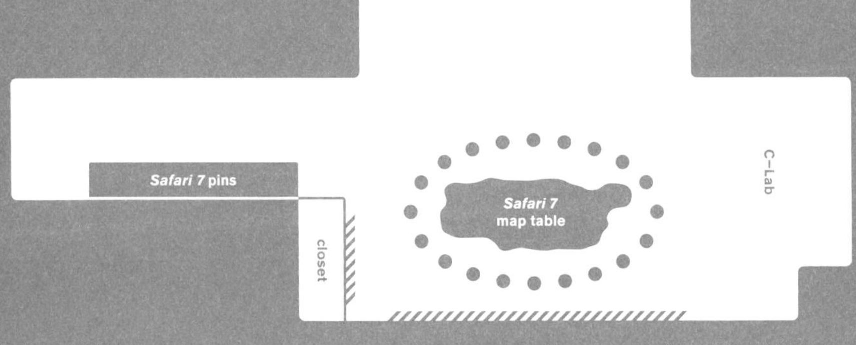

Safari 7 Reading Room Educators' Roundtable
11.18.09
Safari 7 pins
closet
Safari 7 map table
C-Lab

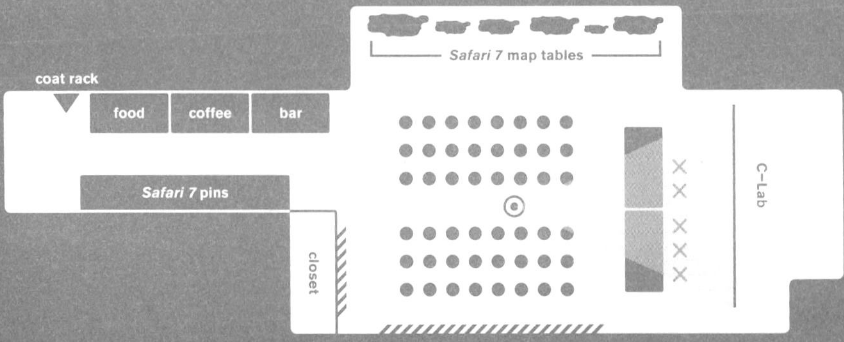

Architecture in Public: A Workshop
12.04–12.05.09
Safari 7 map tables
coat rack
food
coffee
bar
Safari 7 pins
closet
C-Lab

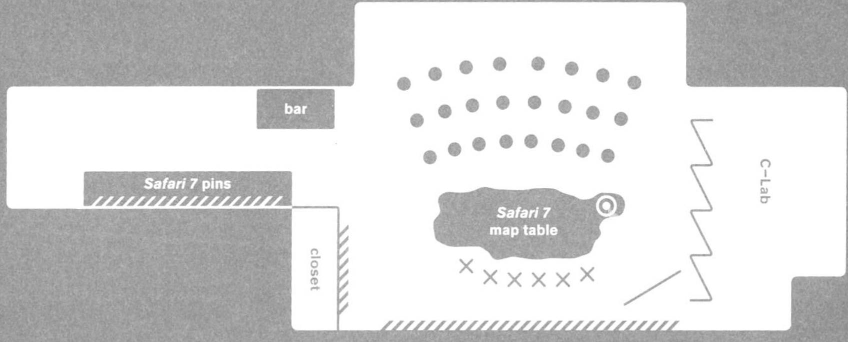

Al Manakh 2: Debate on Tour
12.12.09
bar
Safari 7 pins
closet
Safari 7
map table
C-Lab

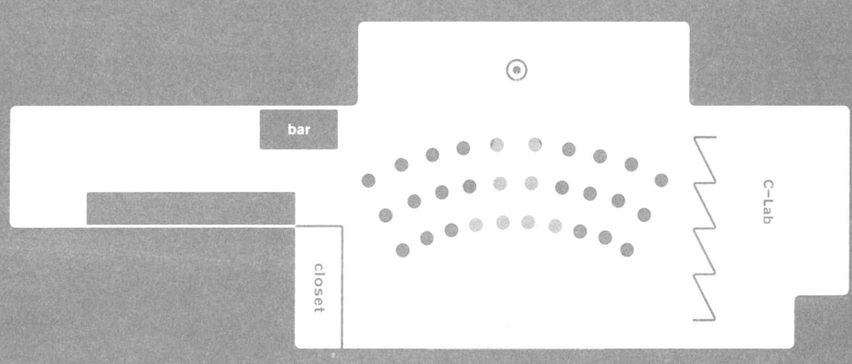

Doctruck at Studio-X: Land and Noise, Space and Silence
01.14.10
bar
closet
C-Lab

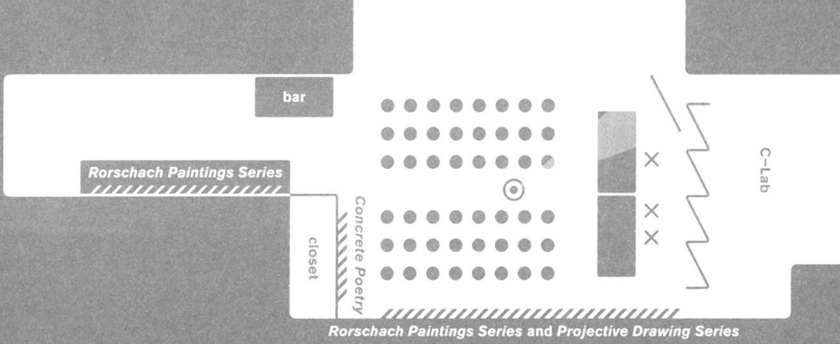

ink Simulcast Conversation
01.28.10
bar
Rorschach Paintings Series
closet
Concrete Poetry
C-Lab
Rorschach Paintings Series and Projective Drawing Series

Discussion on Networked Publics #1: CULTURE
02.09.10

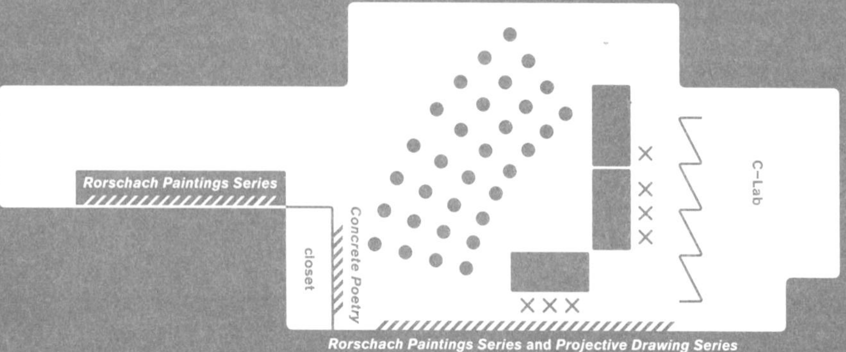

Ink Workshop with Karen Finley and Michelle Fornabai
02.16.10

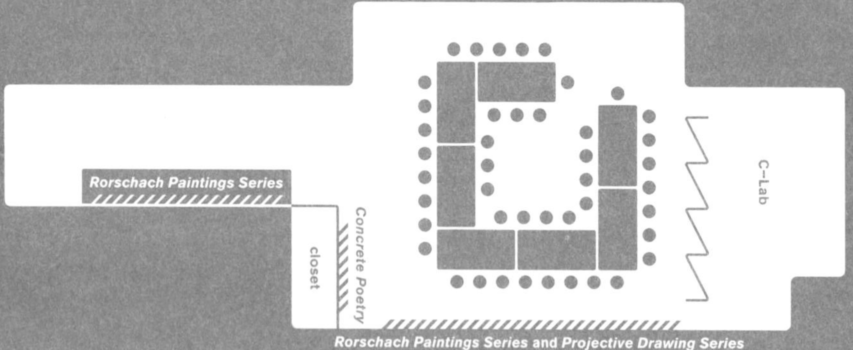

Foodprint NYC
02.27.10

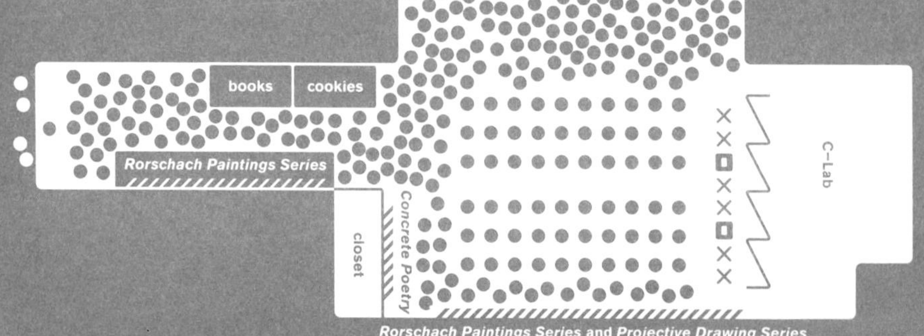

<u>Sustainable Urbanism International</u>
03.11.10
bar
Rorschach Paintings Series
closet
Concrete Poetry
C-Lab
Rorschach Paintings Series and Projective Drawing Series

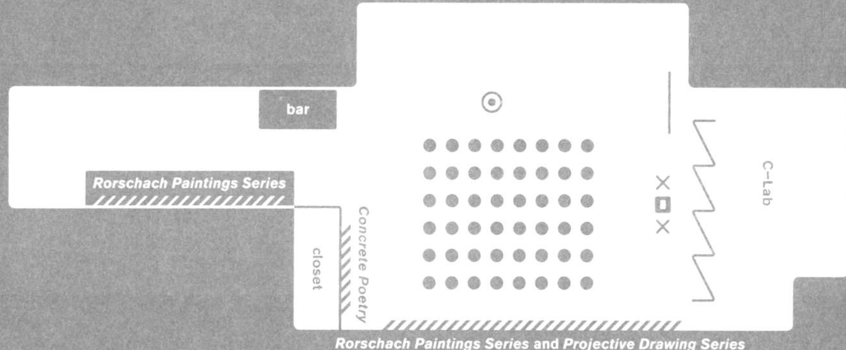

<u>SUPERFRONT at Studio-X</u>
04.01.10
bar
TRANS SIBERIA
closet
C-Lab

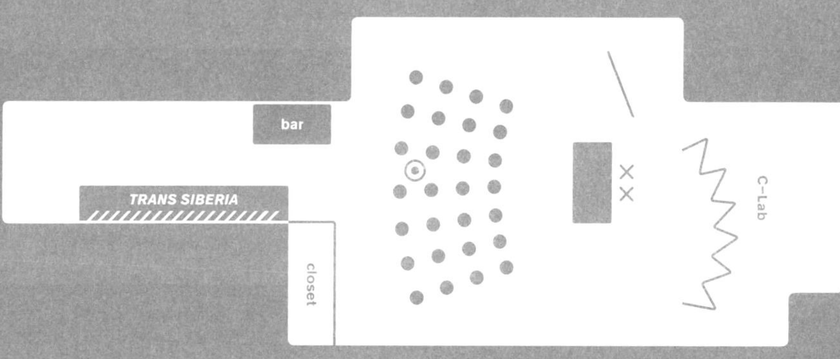

<u>IMAGINATION VESSELS</u> Opening
04.22–05.06.10
bar bar bar
closet
C-Lab
DJ
IMAGINATION VESSELS

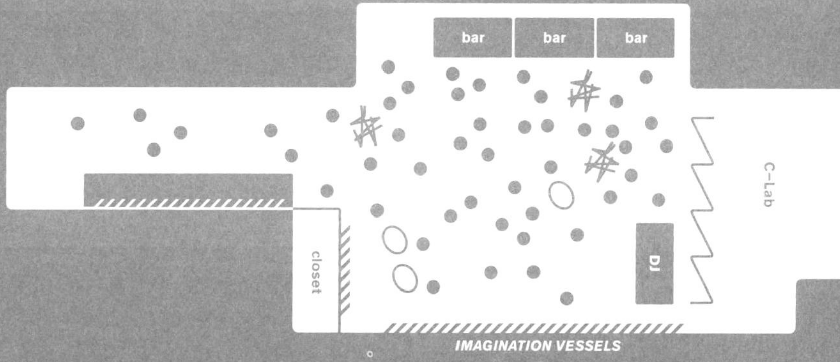

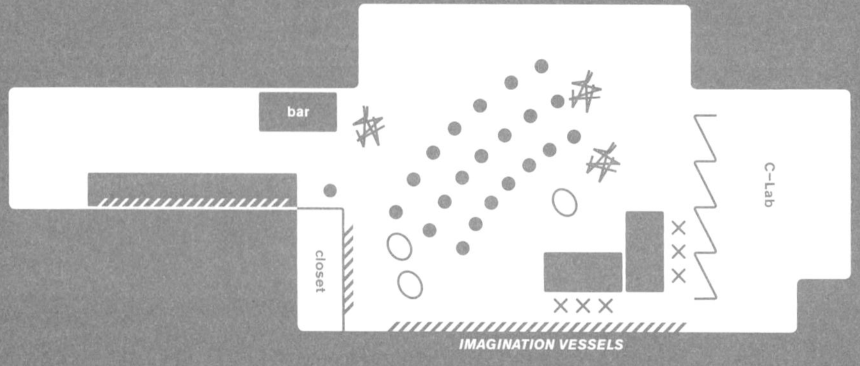

bar
closet
C-Lab
IMAGINATION VESSELS

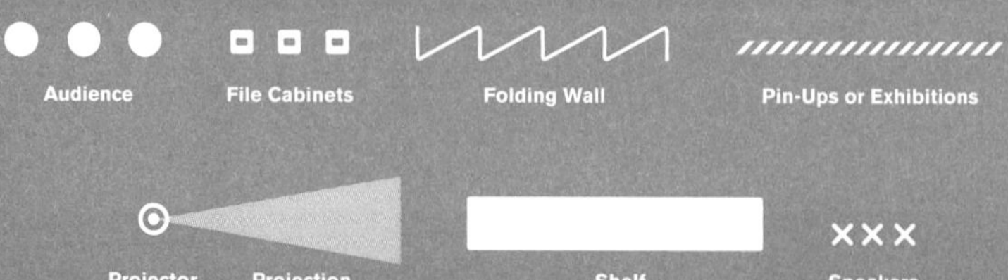

Audience
File Cabinets
Folding Wall
Pin-Ups or Exhibitions
Projector
Projection
Shelf
Speakers

**Populate.
People make a place vital. These
are the ones who have done that
for Studio-X New York.**

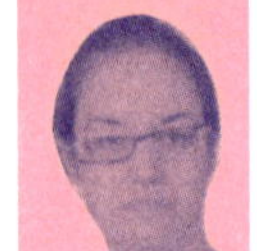

Magda Aboulfadl
Manhattan
Community
Board 5

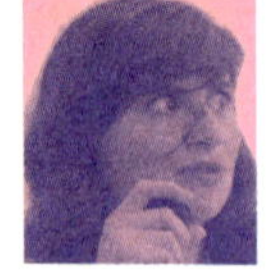

Amale Andraos
WORKac

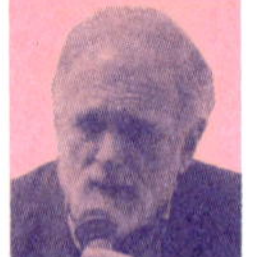

Moshe Adler
Columbia GSAPP

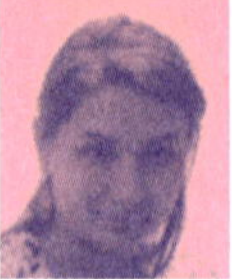

Erin Aigner
New York Times

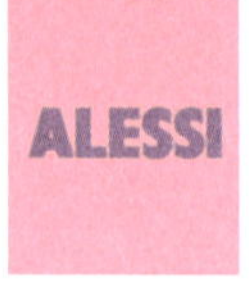

Alessi

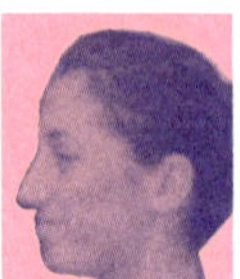

Adrien Allred
Pratt Institute

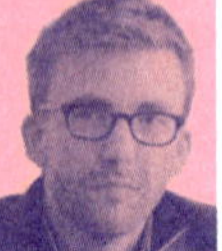

Nora Arkawi
Columbia GSAPP

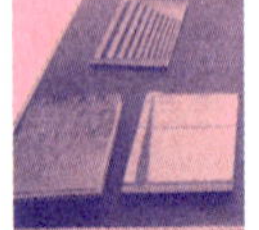

Tobias Armborst
Interboro

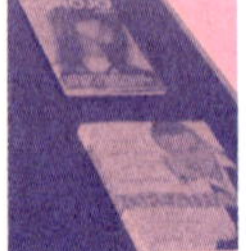

ARO

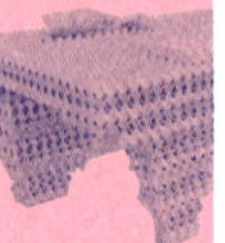

ARROJO Studio

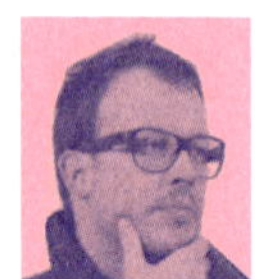

Ate Atema
Atema
Architecture

Markus Bader
Raumlabor

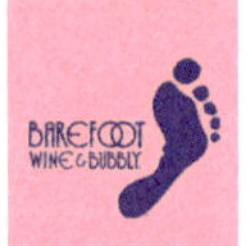

Barefoot Wine

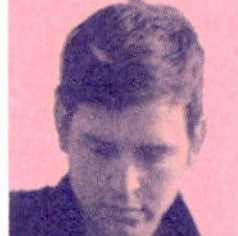

Christopher Barley
Columbia GSAPP

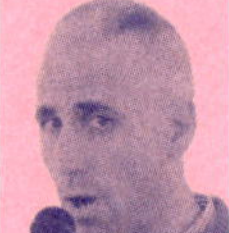

Sean Basinski
Street Vendor
Project

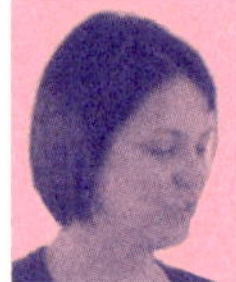

Julie Behrens
Columbia GSAPP

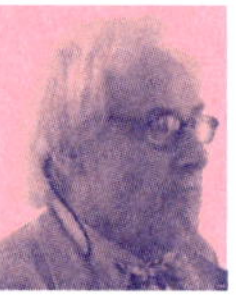

Byron Bell
Bell Larson
Architects and
Planners

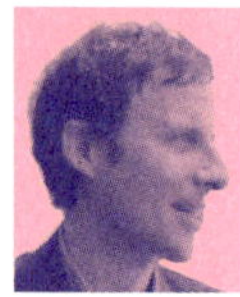

David Benjamin
Living
Architecture Lab
Columbia GSAPP

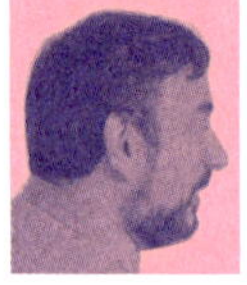

Joel Berg
New York City
Coalition Against
Hunger

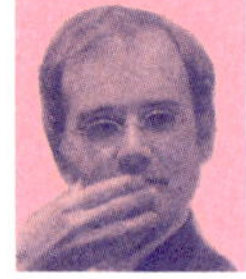

Barry Bergdoll
Columbia and
MoMA

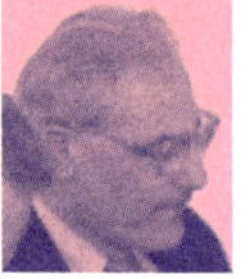

Aaron Betsky
The Cincinnati Art
Museum

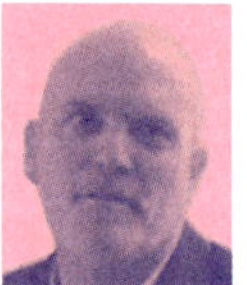

Matthew Bialecki
Bialecki
Architects

Matthew Bissen
Parsons The New
School for Design

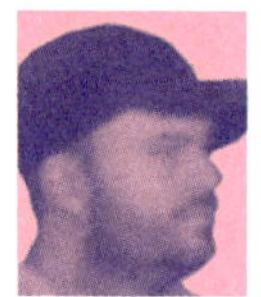

Jacob Blak
UiWE

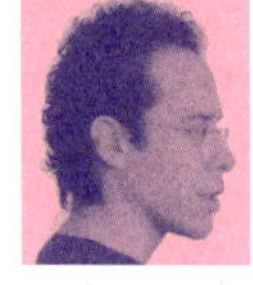

Jonathan Bogarín
artist

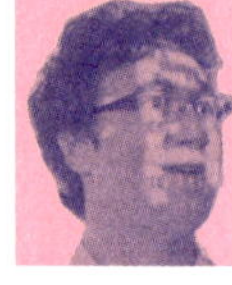

Alfredo
Brillembourg
S.L.U.M. LAB
Columbia GSAPP

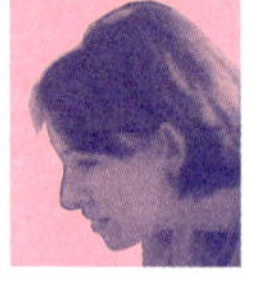

Marlo Brown
Columbia GSAPP

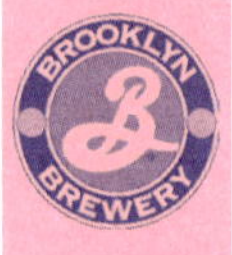

Brooklyn Brewery

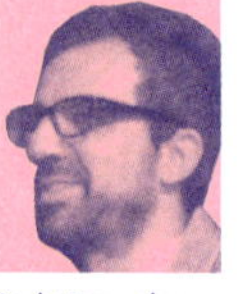

Gavin Browning
Studio-X New York

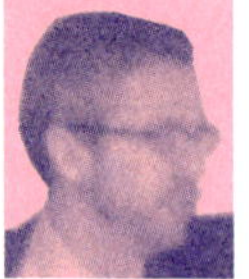

Daniel Buezna
Columbia School
of Business

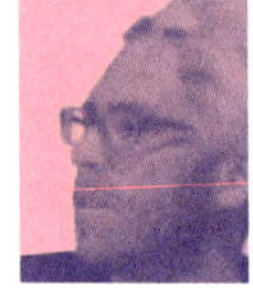

Craig Buckley
Columbia GSAPP

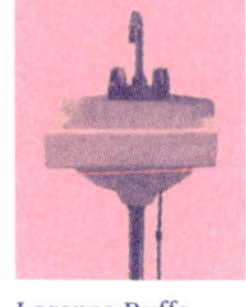

Lorenzo Buffa
industrial
designer

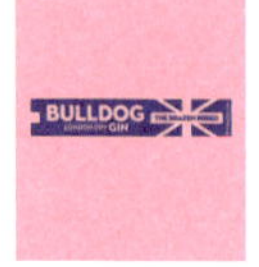

Bulldog Gin

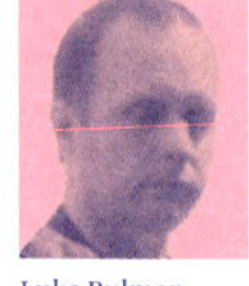

Luke Bulman
Thumb

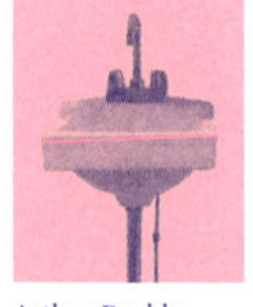

Arthur Burkle
actor

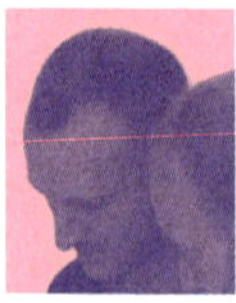

Felix Burrichter
PIN–UP

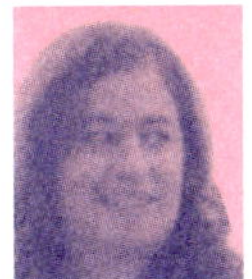

Greta Byrum
Nobody Books

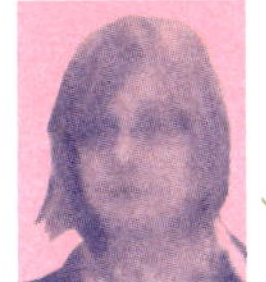

Heather Cardinale
NYC Dept of
Education

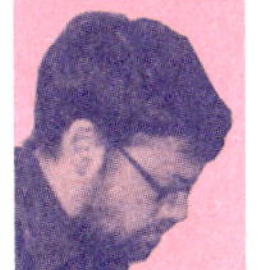

Hrolfur Karl Cela
Parsons The New
School for Design

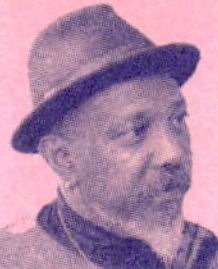

Sergio Cezar
artist

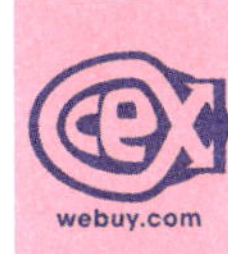

CEX Complete
Entertainment
Exchange

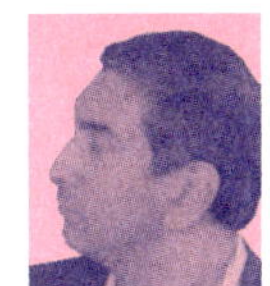

Vishaan
Chakrabarti
Columbia GSAPP

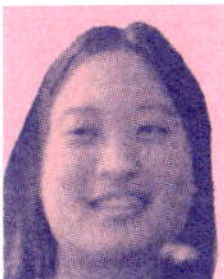

Olivia Chen
Inhabitat

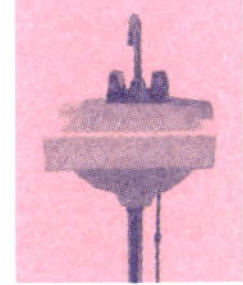

Tom Chiu
FLUX Quartet

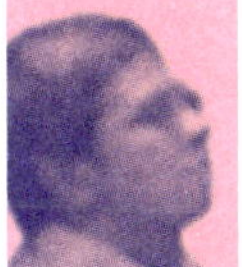

Eileen Clancy
I-Witness Video

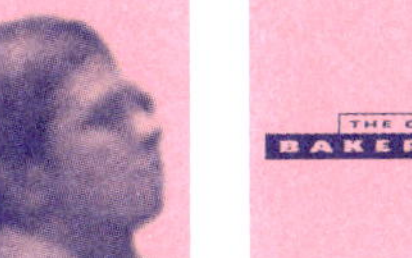

The City Bakery

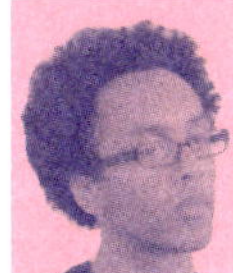

Jace Clayton
(DJ /rupture)
musician

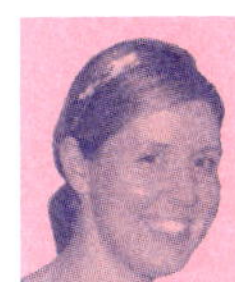

Sarah Cloonan
Columbia GSAPP

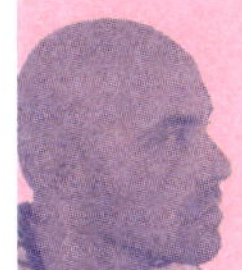

Benedict Clouette
C-Lab
Columbia GSAPP

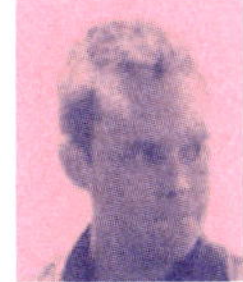

Zachary Colbert
PRE-Office

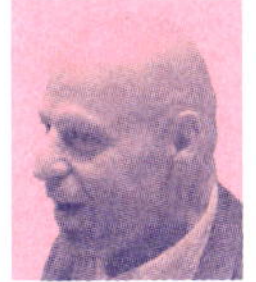

Jean-Louis Cohen
NYU

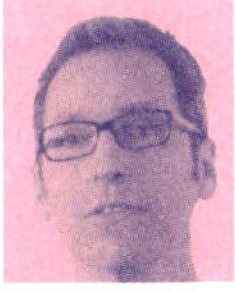

Nevin Cohen
The New School
University

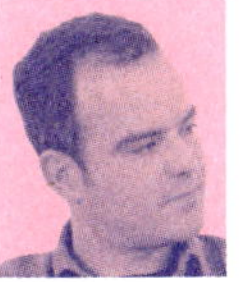

Marcelo Coelho
MIT Media Lab

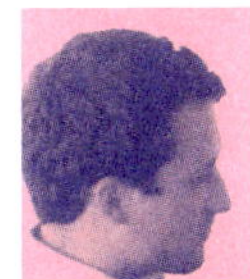

Wayne Congar
Columbia GSAPP

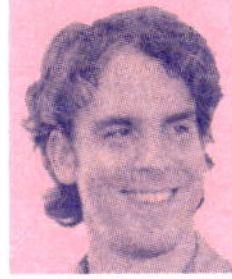

Raul Correa-Smith
Columbia GSAPP

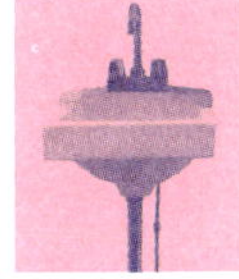

Rachel Crumpler
Queens Museum
of Art

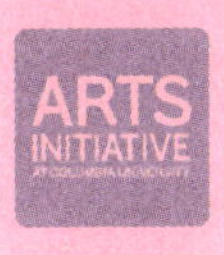

CU Arts Initiative

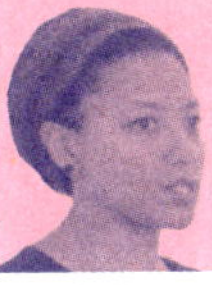

Makalè Faber
Cullen
Slow Food

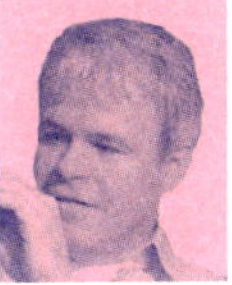

Glen Cummings
MTWTF

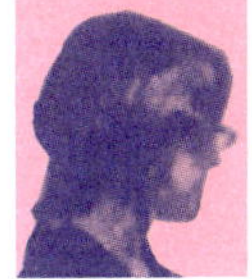

Elizabeth Currid
USC

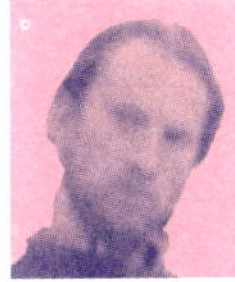

Nathaniel Curtis
iLAND Art

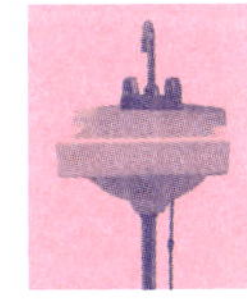

Sarah Dahlen
actor

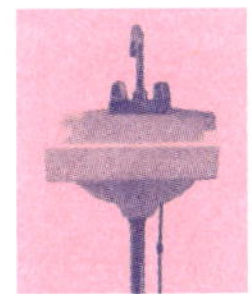

Amanda Dargan
City Lore

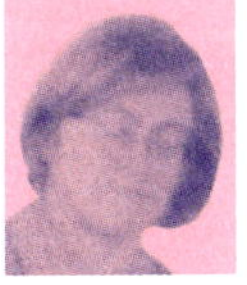

Cynthia Davidson
Log

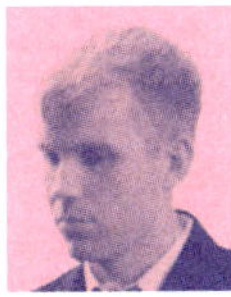

Aaron Davis
PRE-Office

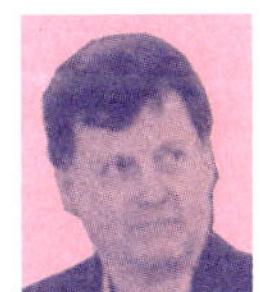

Alex Deffner
University of
Thessaly

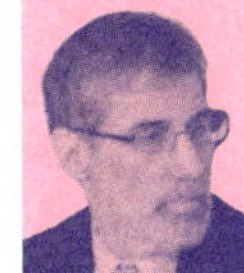

Barry Dinerstein
NYC Dept of City
Planning

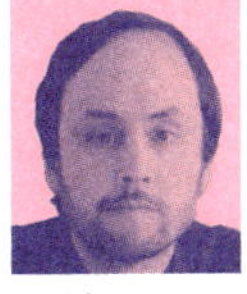

Daniel D'Oca
Interboro

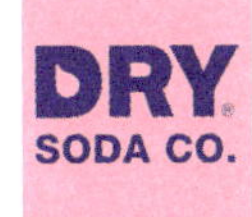

Dry Soda Co.

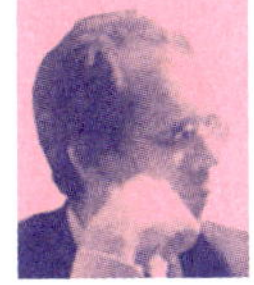

Stephen
Duncombe
NYU

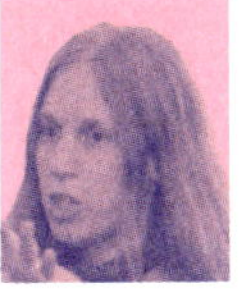

Nora Libertun de
Duren
NYC Dept of Parks
and Recreation

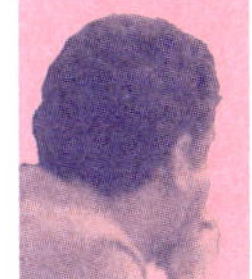

Edward Eckert
Growing Up
Green
Charter School

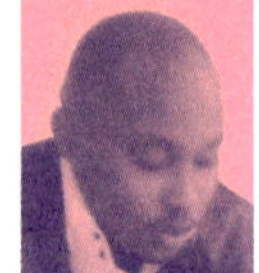

Yern Edozie
Columbia GSAPP

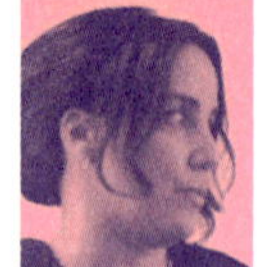

Solange Fabião
architect

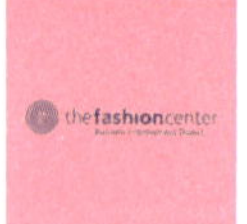

Fashion Center
Business
Improvement
District

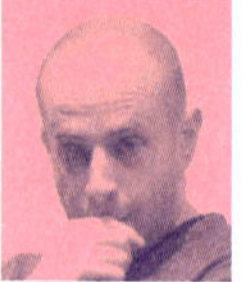

Yevgeniy Fiks
artist

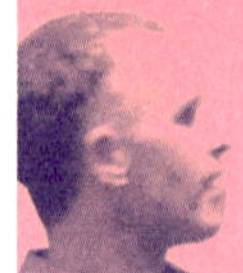

Micah Fink
filmmaker

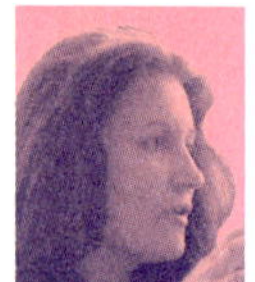

Karen Finley
artist

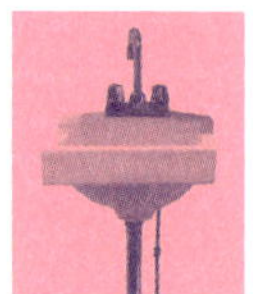

David First
musician

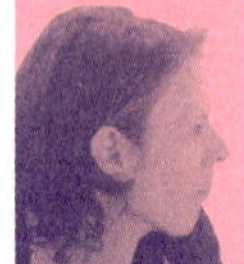

Lars Fischer
common room

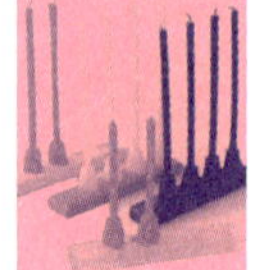

Kegan Fisher
Design Glut

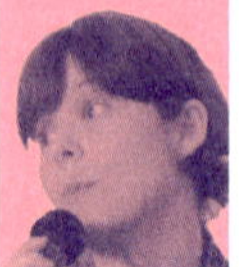

Rebecca Federman
New York Public
Library

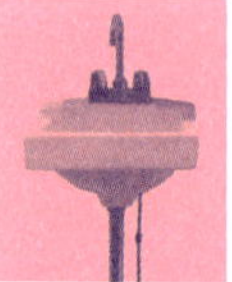

Sierra Feldner-
Shaw
writer

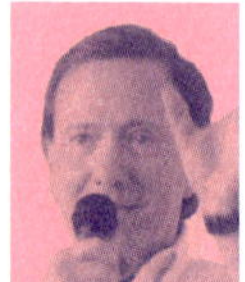

Stanley Fleishman
Jetro Cash and
Carry/ Restaurant
Depot

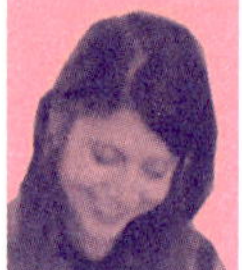

Michelle Fornabai
Columbia GSAPP

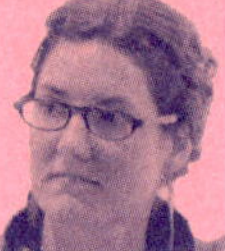

Anne Frederick
Hester Street
Collaborative

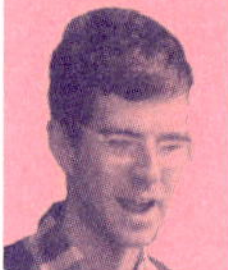

Steven Garcia
Columbia GSAPP

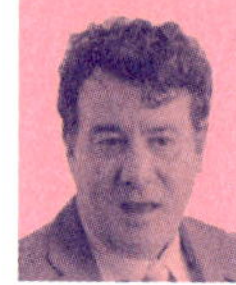

Douglas Gauthier
Gauthier
Architects

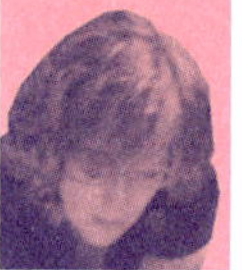

Renee Glick
Pratt Institute

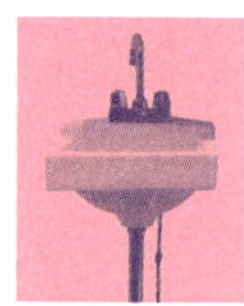

Cristina Goberna
Fake Industries
Architectural
Antagonism

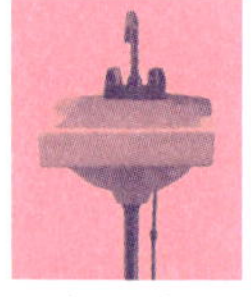

Stephen Graham
University of
Newcastle

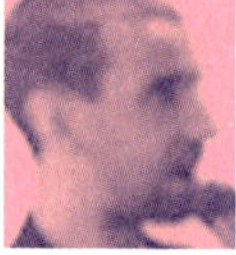

Joseph Grima
Domus

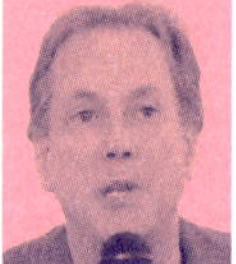

William Grimes
New York Times

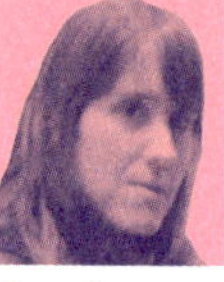

Laura Hanna
filmmaker

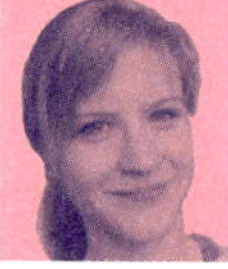

Greta Hansen
Warm Engine

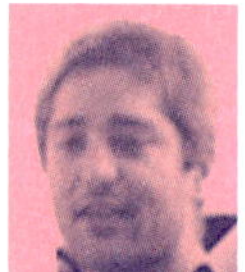

Usman Haque
haque :: design +
research

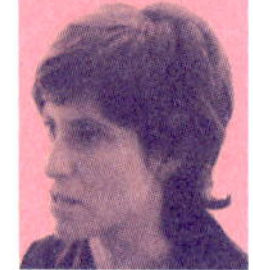

Larissa Harris
Queens Museum
of Art

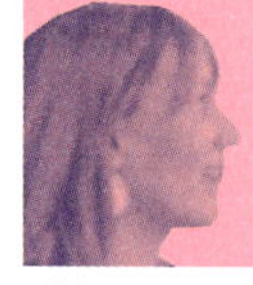

Leigh Harvey
Fashion Center
Business Improve-
ment District

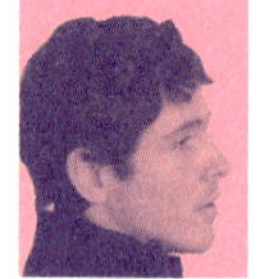

David Haskell
New York Magazine
and Kings County
Distillery

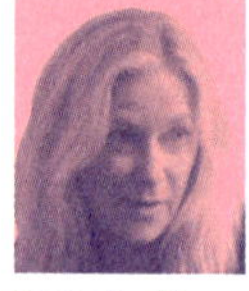

Laurie Hawkinson
Columbia GSAPP

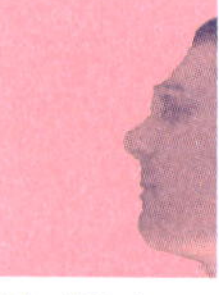

Waleed Hazbun
The John Hopkins
University

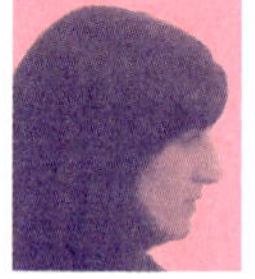

Sarah Herda
Graham
Foundation

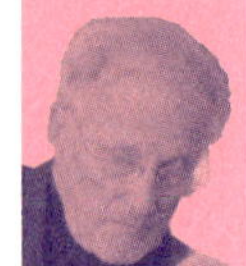

Stan Herman
Council of Fashion
Designers of
America

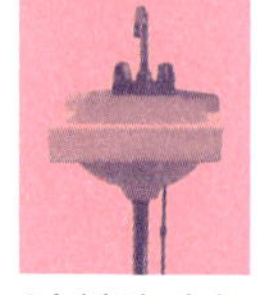

Gabriel Hirnthaler
Villa Feuerlöscher

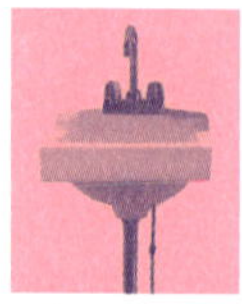

Barbara Hirnthaler
Villa Feuerlöscher

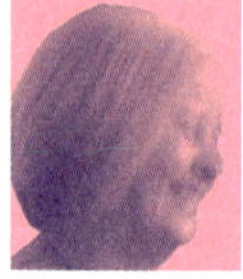

Brooke Hodge
curator

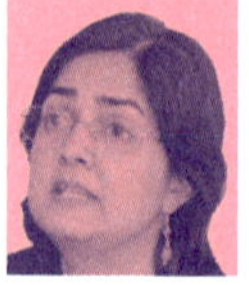

Jyoti Hosagrahar
Columbia GSAPP
and SIPA

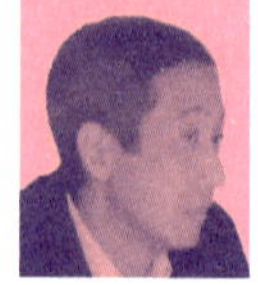

Jeffrey Inaba
C-Lab
Columbia GSAPP

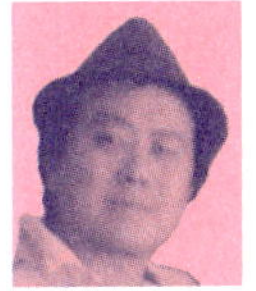

Andrew Ingkavet
composer

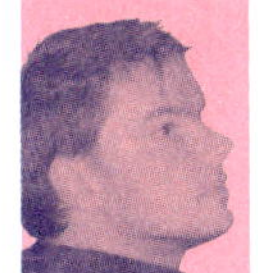

Bjarke Ingels
BIG

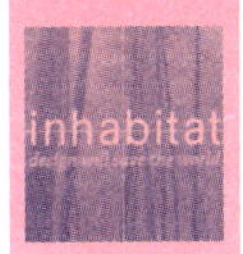

Inhabitat

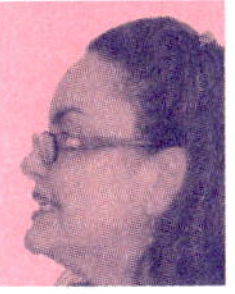

Clara Irazábal
Columbia GSAPP

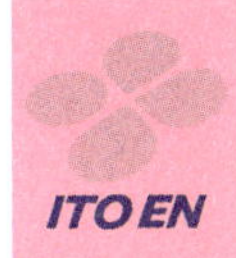

Ito En

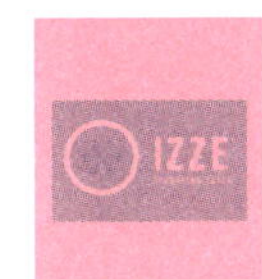

IZZE Sparkling
Juice

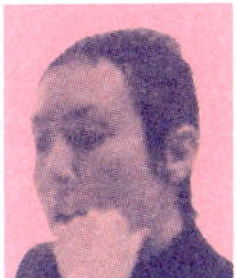

Marisa Jahn
artist

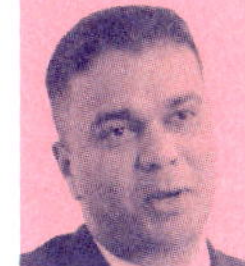

Jack Jaskaran
NYPD

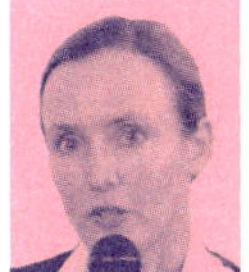

Natalie
Jeremijenko
NYU

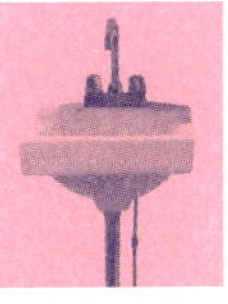

David Jensensius
spurse

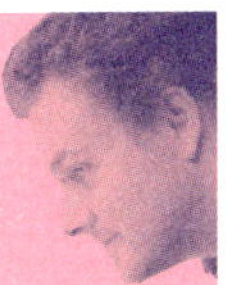

Mitchell Joachim
Terreform ONE

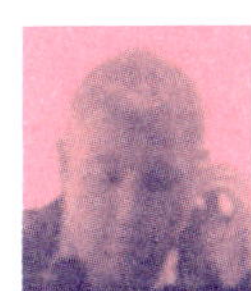

Jeffrey Johnson
China Lab
Columbia GSAPP

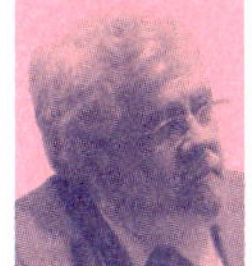

David Cay
Johnston
writer

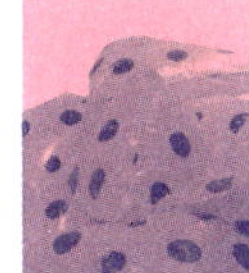

Adrian
Kondratowicz
artist

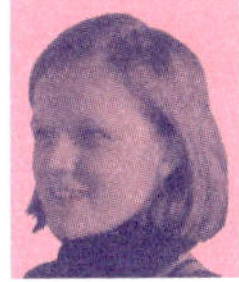

Dana Karwas
C-Lab
Columbia GSAPP

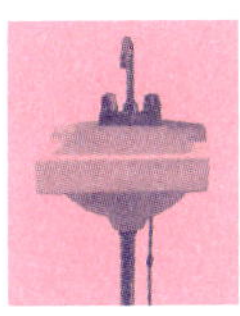

Chris Kao
architect

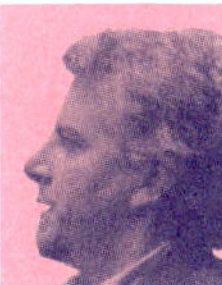

Keith Kaseman
Columbia GSAPP

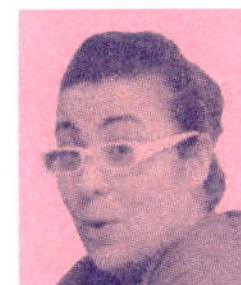

Olympia Kazi
Van Alen Institute

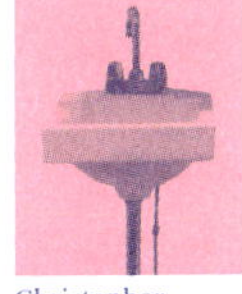

Christopher
Kennedy
educator

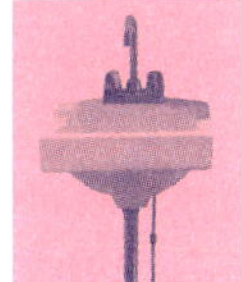

Anna Kenoff
Columbia GSAPP

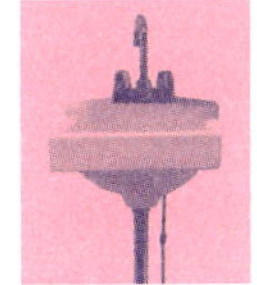

Iain Kerr
spurse

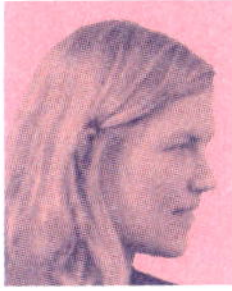

Kate Kerrigan
DUMBO
Improvement
District

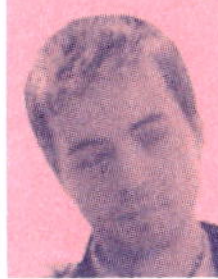

Daniel Kidd
PRE-Office

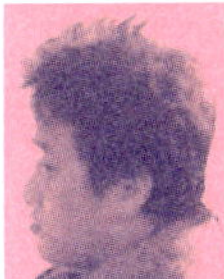

Kyung Jae Kim
Columbia GSAPP

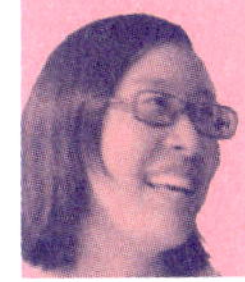

Janette Kim
Urban Landscape
Lab
Columbia GSAPP

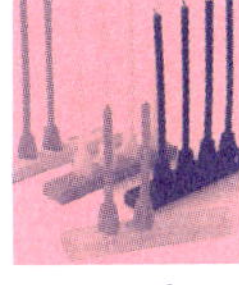

Liz Kinnmark
Design Glut

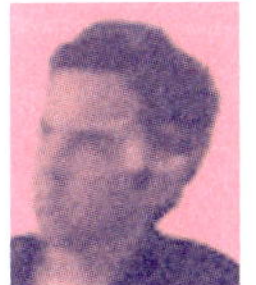

Hubert Klumpner
S.L.U.M. LAB
Columbia GSAPP

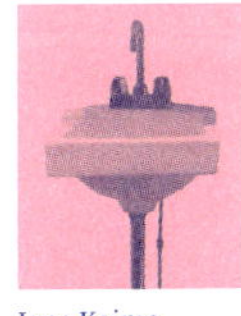

Jane Kojma
DUMBO
Improvement
District

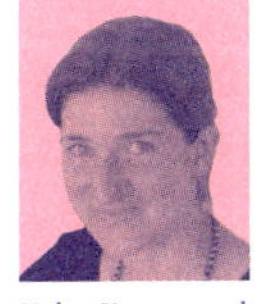

Helen Kongsgaard
Urban Landscape
Lab
Columbia GSAPP

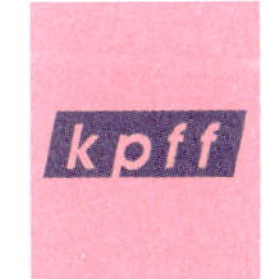

KPFF Consulting
Engineers

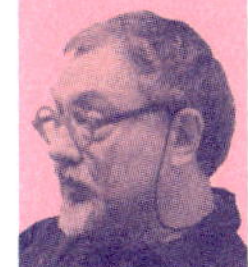

Vitaly Komar
artist

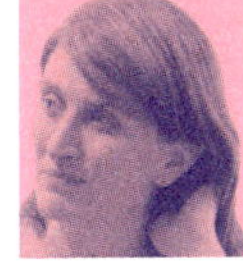

Kaja Kühl
Columbia GSAPP

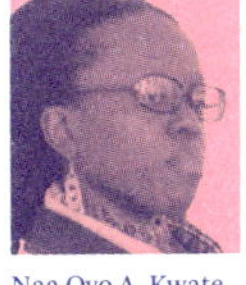

Naa Oyo A. Kwate
Columbia
Mailman School
of Public Health

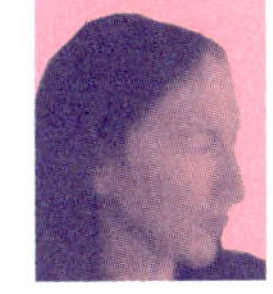

Sylvia Lavin
UCLA

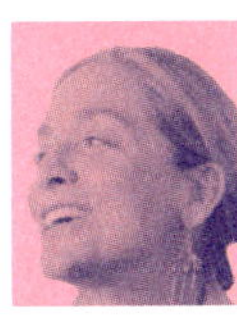

Annie Hauck-
Lawson
Brooklyn College

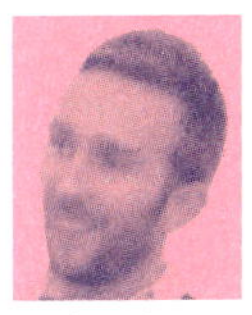

David van der Leer
Guggenheim
Museum

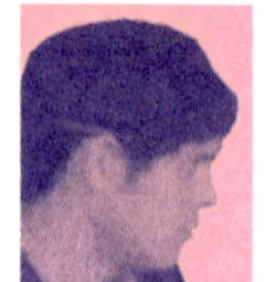
Christopher Leong
Leong Leong
Architecture

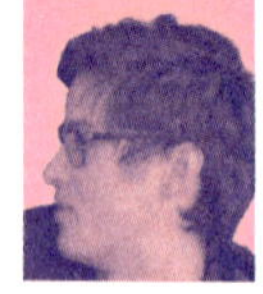
Dominic Leong
Leong Leong
Architecture

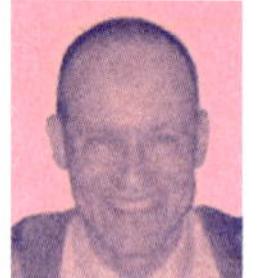
Andres Lepik
MoMA

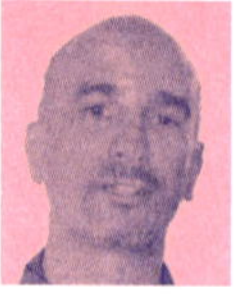
Frederic Levrat
Columbia GSAPP

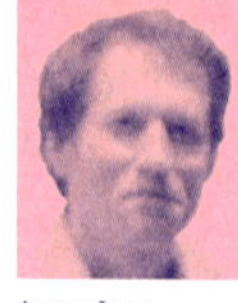
Aaron Levy
Slought
Foundation

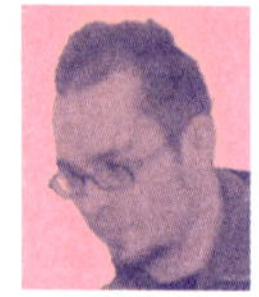
Christos Liorius
University of
Thessaly

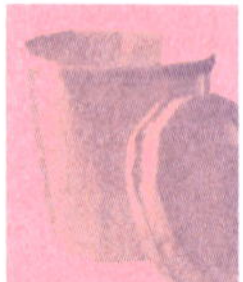
Paul Loebach
furniture designer

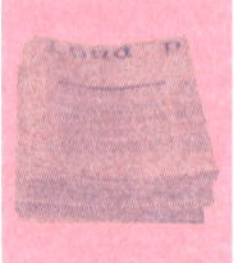
loud paper

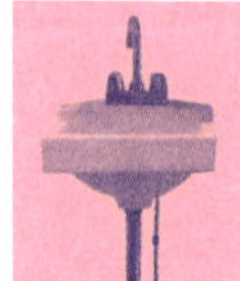
Brian Loughlin
Jersey City
Housing Authority

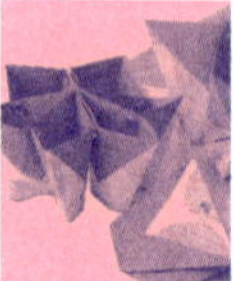
Melissa MacNair
industrial
designer

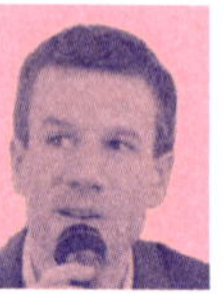
Geoff Manaugh
BLDGBLOG

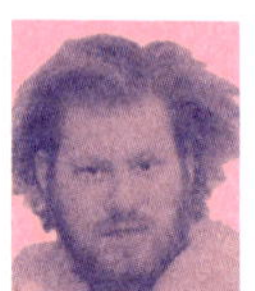
Michael
Mandiberg
artist

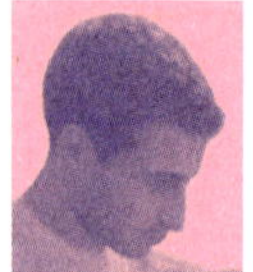
Hart Marlow
Pratt Institute

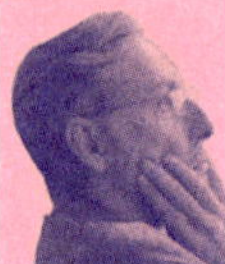
Reinhold Martin
Columbia GSAPP

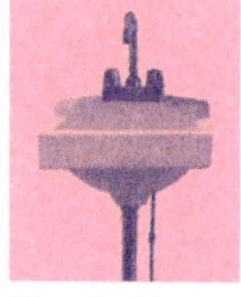
Diana Martinez
Columbia GSAPP

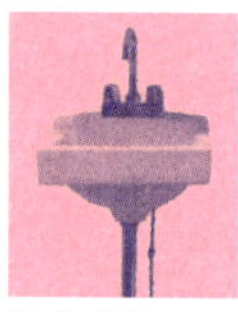
Christof Mayer
Raumlabor

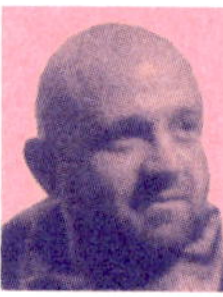
Jürgen Mayer H.
J. MAYER H.

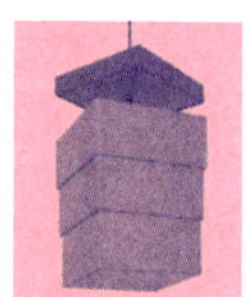
Kevin Patrick
McCarthy
KPM light and
shade

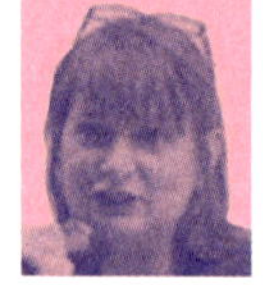
Amanda
McDonald Crowley
Eyebeam

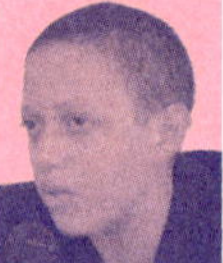
Mitch McEwen
SUPERFRONT

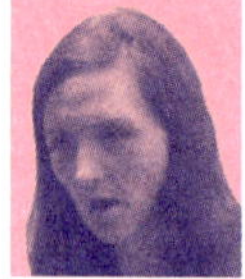
Kathrin McInnis
filmmaker

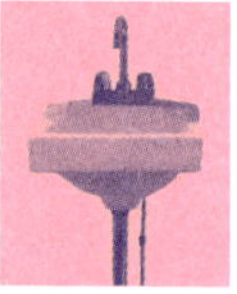
Katharine Meagher
C-Lab
Columbia GSAPP

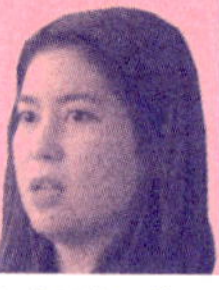
Leah Meisterlin
PRE-Office

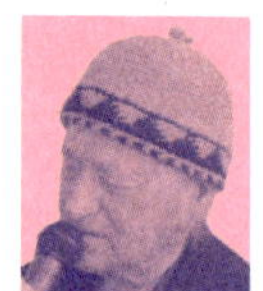
Jonas Mekas
Anthology Film
Archives

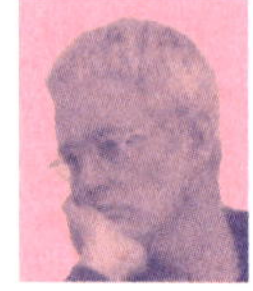
William Menking
*The Architect's
Newspaper*

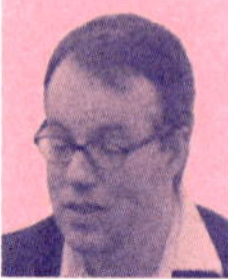
Michael Meredith
MOS

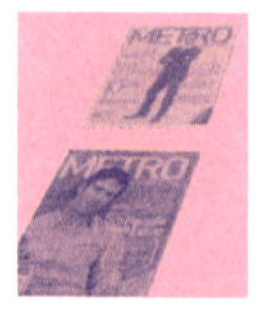
*Metrosource
Magazine*

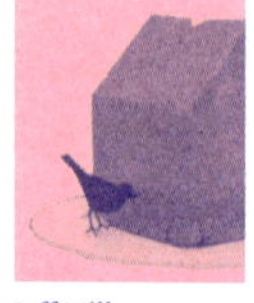
Jeff Miller
industrial designer

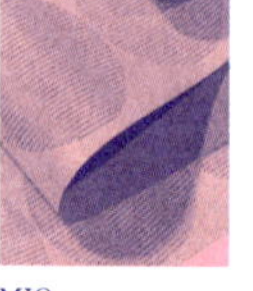
MIO
furniture designer

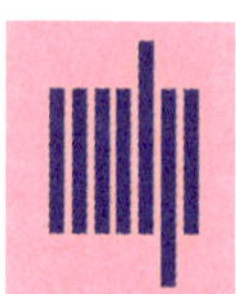
The MIT Press

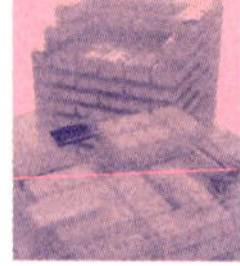
Takeshi Miyakawa
Takeshi Miyakawa
Design

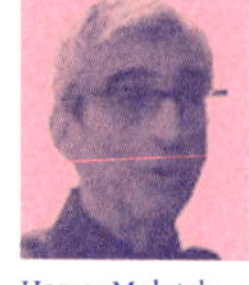
Harvey Molotch
NYU

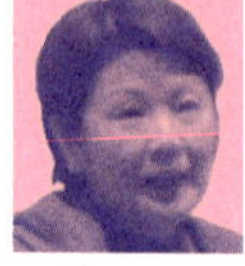
Toshiko Mori
Toshiko Mori
Architect

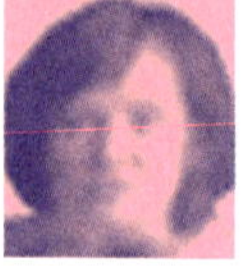
Petia Morozov
Columbia GSAPP

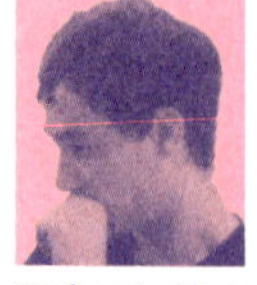
Stephen Mosblech
Nobody Books

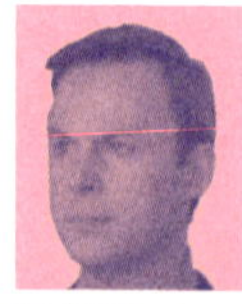
Patrick Murphy
NYC Economic
Development
Corporation

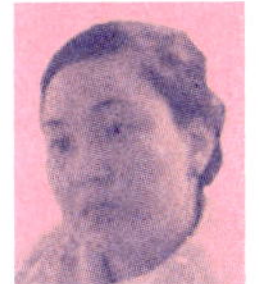

Daisy Nam
Columbia School
of the Arts

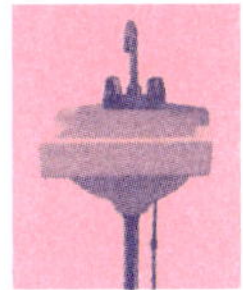

Deborah Natsios
Cryptome

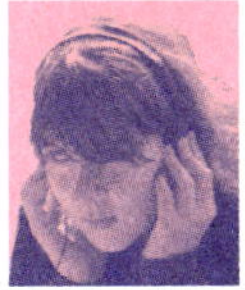

Minna Ninova
SIDL
Columbia GSAPP

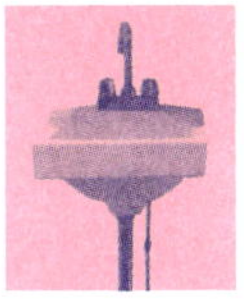

Matthias Neumann
architect

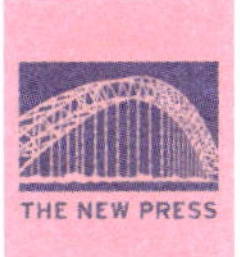

The New Press

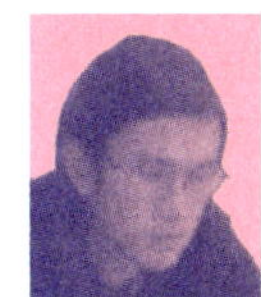

Thad Nobuhara
Columbia GSAPP

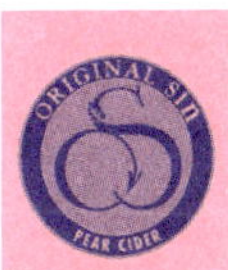

Original Sin Hard
Cider

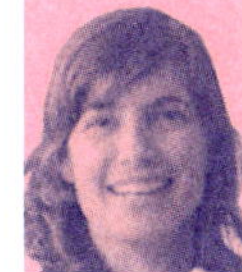

Kate Orff
Urban Landscape
Lab
Columbia GSAPP

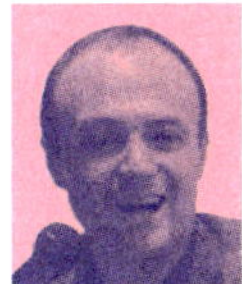

Jorge Otero-Pailos
Columbia GSAPP

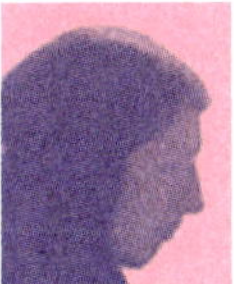

Margot Otten
Parsons The New
School for Design

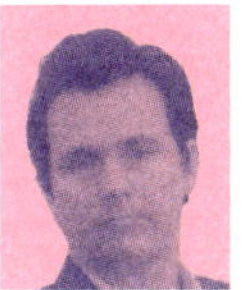

Nicolai Ouroussoff
New York Times

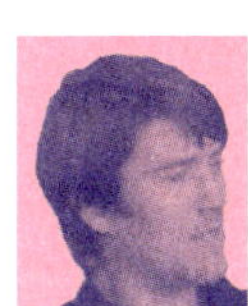

Christian Pagh
UiWE

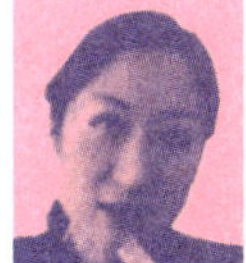

Qing Pan
National Art
Museum of China

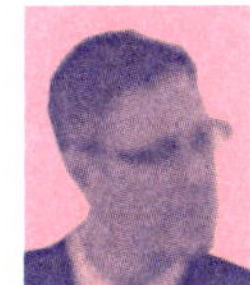

Frank Pasquale
Seton Hall Law
School

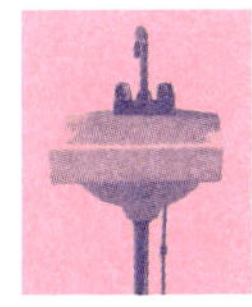

Jonathan Payne
Columbia GSAPP

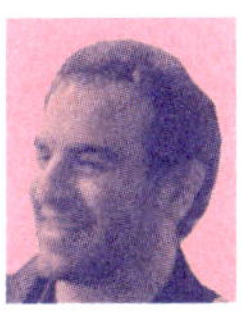

Daniel Perlin
artist

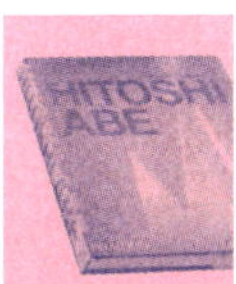

Phaidon Press

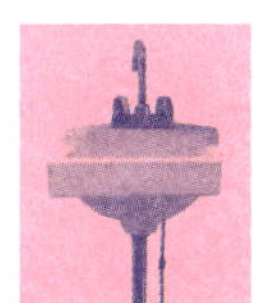

Hannes Preisch
artist

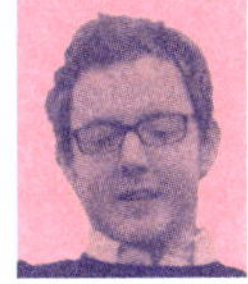

Will Prince
PARK

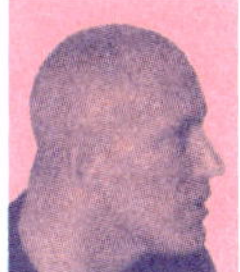

Joshua Prince-
Ramus
REX

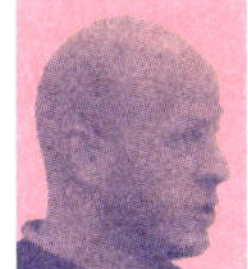

Benjamin Prosky
Columbia GSAPP

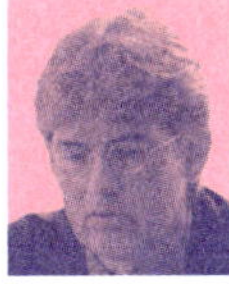

Todd Rader
Rader + Crews

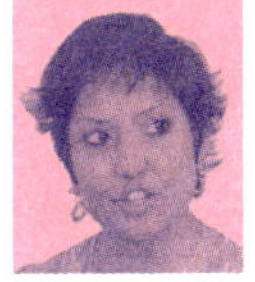

Suman
Raghunathan
Progressive States
Network

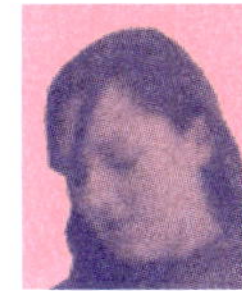

Rachael Rakes
DocTruck

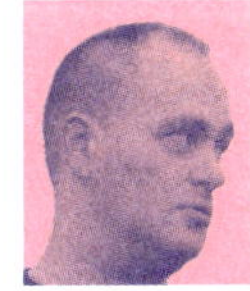

Enrique Ramirez
a456

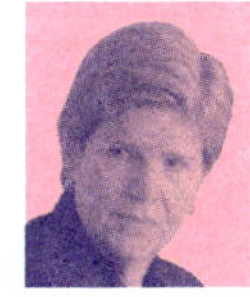

Barbara Randall
Fashion Center
BID

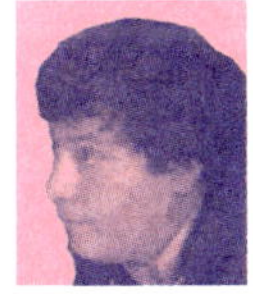

Perry Randazzo
CCNY

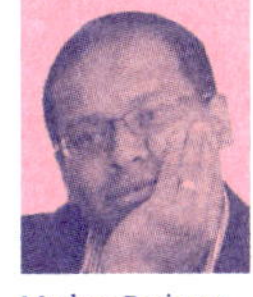

Mathan Ratinam
MILC

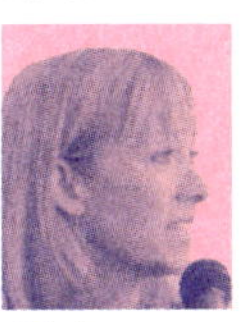

Christina Ray
CHRISTINA RAY

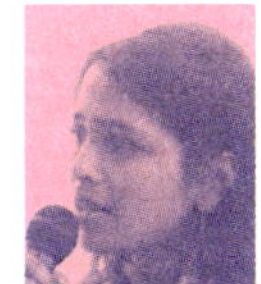

Prerana Reddy
Queens Museum
of Art

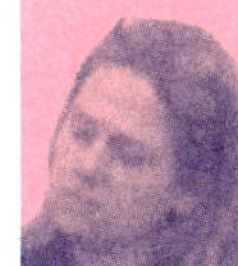

Amanda Reeser
Lawrence
Praxis

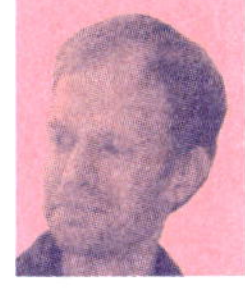

David Reinfurt
Columbia GSAPP

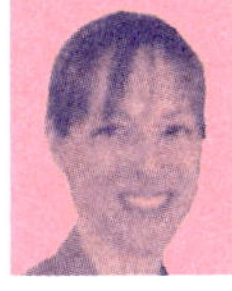

Jen Renzi
writer

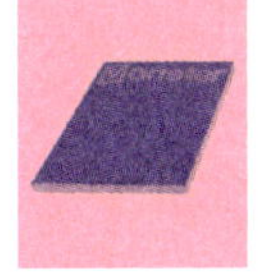

REX

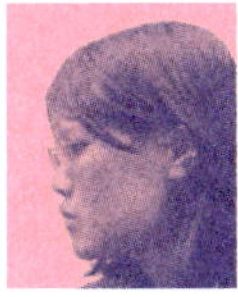

Pollyanna Rhee
Columbia GSAPP

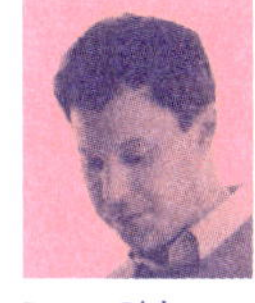

Damon Rich
Center for Urban
Pedagogy

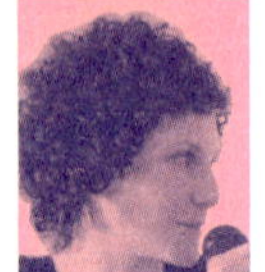

Sarah Rich
writer

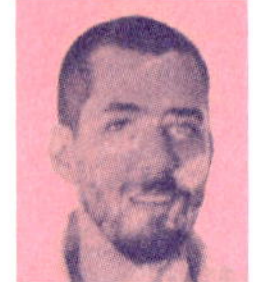

Pedro Rivera
Studio-X Rio

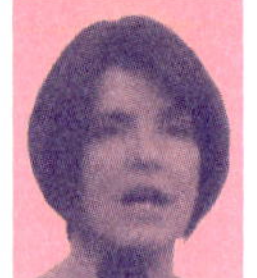

Lisa Rochon
Toronto Globe and Mail

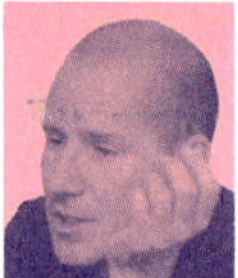

Michael Rock
2x4

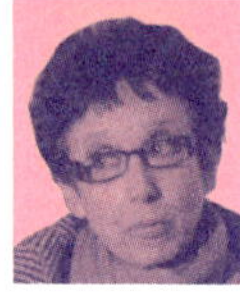

Avital Ronell
NYU

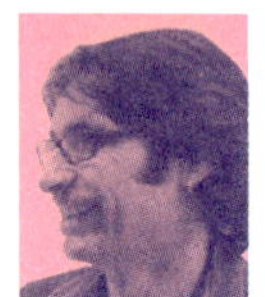

Andrew Ross
NYU

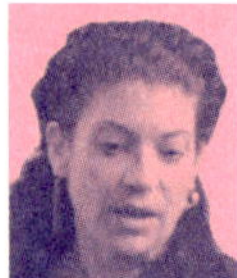

Karla Rothstein
Columbia GSAPP

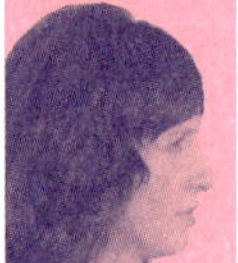

Heather Rowe
artist

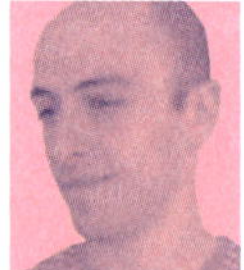

Dan Rubinstein
Surface

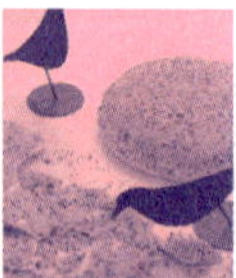

Andrea Ruggiero
industrial designer

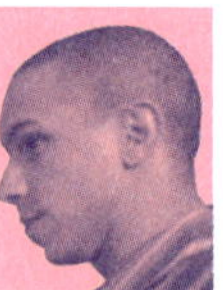

Orlando Rymer
CCNY

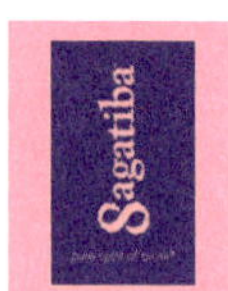

Sagatiba

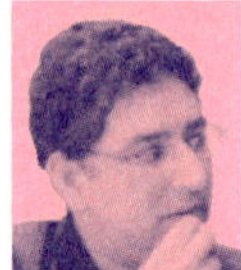

Sukhdev Sandhu
NYU and *Daily Telegraph*

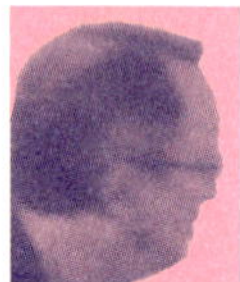

William Saunders
Harvard Design Magazine

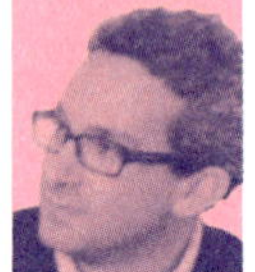

David Sax
writer

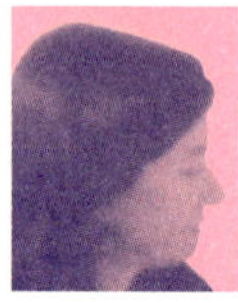

Ashley Schafer
Praxis

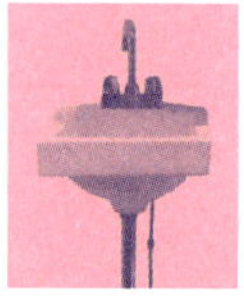

Brett Seamans
CCNY

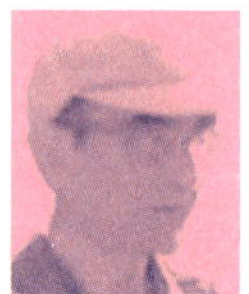

Raja Shehadeh
writer

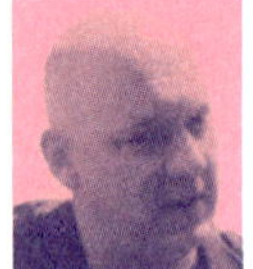

Mark Shepard
University at Buffalo

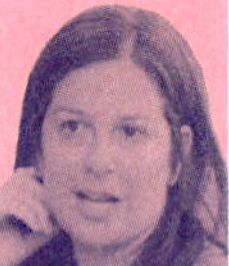

Brooke Singer
artist

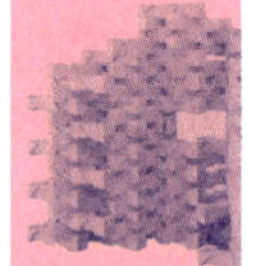

Situ Studio

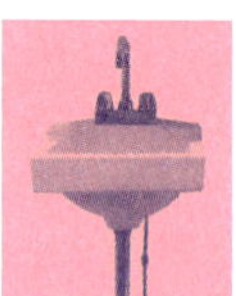

Jay Smith
actor

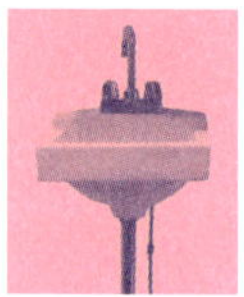

William Brian Smith
Columbia GSAPP

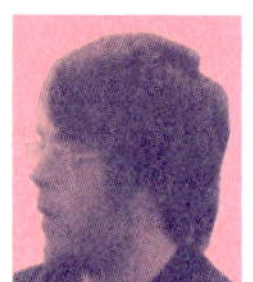

Danny Snelson
Nobody Books

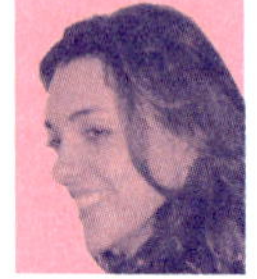

Federica Soletta
Columbia GSAPP

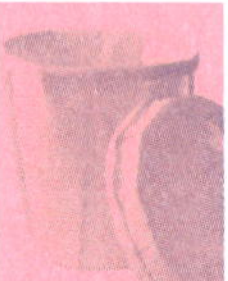

Christopher Specce
industrial designer

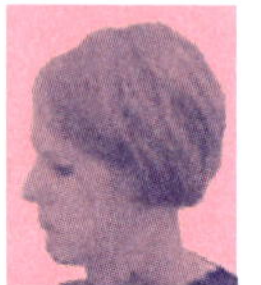

Molly Wright Steenson
Princeton University

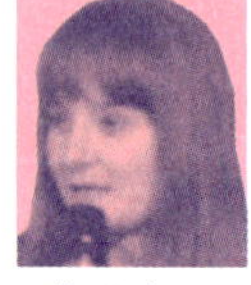

Halina Steiner
CCNY

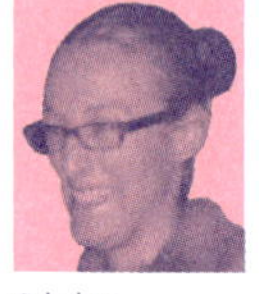

Cristina Steingräber
Hatje Cantz Verlag

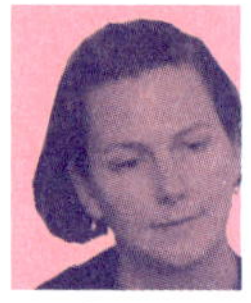

Georgianna Stout
2x4

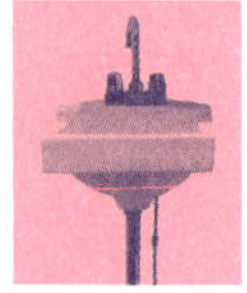

Abbot Street
reader

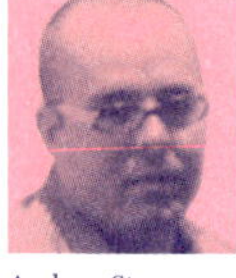

Andrew Sturm
curator

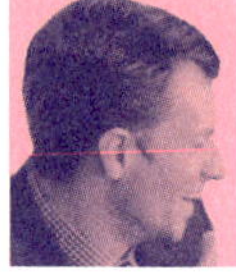

Robert Sumrell
AUDC

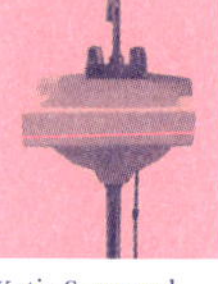

Katja Szymczak
Raumlabor

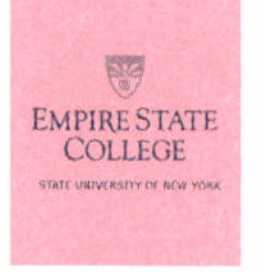

SUNY Empire State College

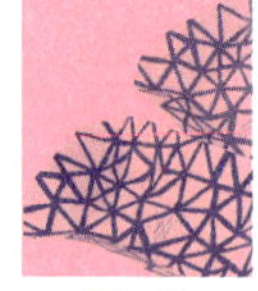

So Takahashi
So Takahashi design studio

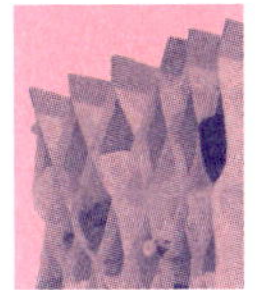

Megan Talley
industrial designer

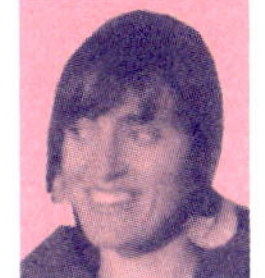

Astra Taylor
filmmaker

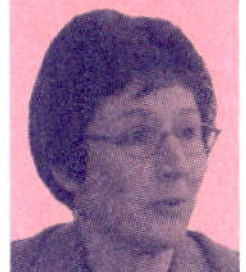

Beverly Tepper
Rutgers University

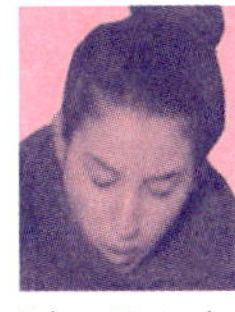

Kristen Teutonico
Parsons The New
School for Design

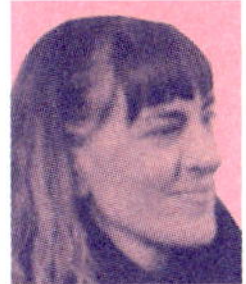

Ioanna Theocharo-
poulou
Parsons The New
School for Design

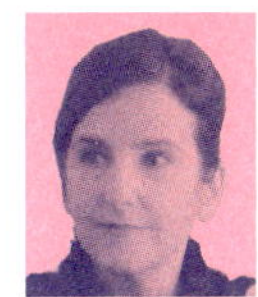

Georgeen
Theodore
Interboro

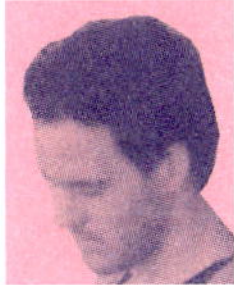

Troy Therrien
Columbia GSAPP

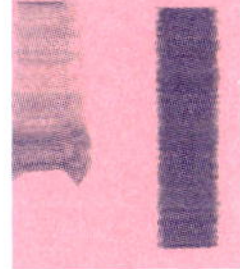

Suzanne Tick
Suzanne Tick, Inc

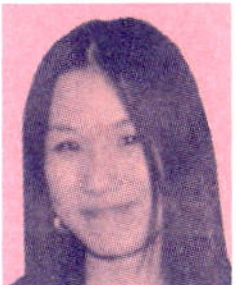

Tong Tong
Columbia GSAPP

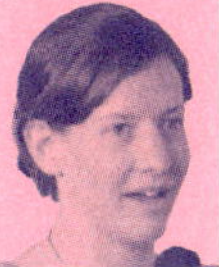

Nicola Twilley
Edible Geography

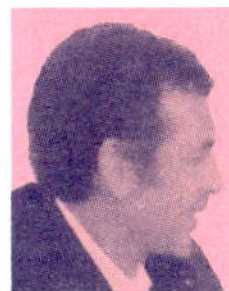

Henry Urbach
SFMOMA

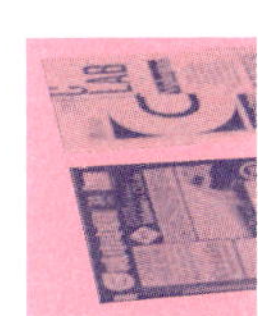

Urban China

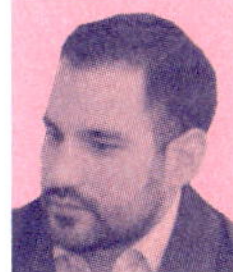

Javier Valdes
Make The Road
New York

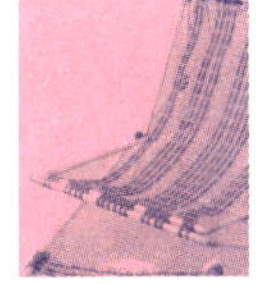

Alex Valich
Redst/Collective

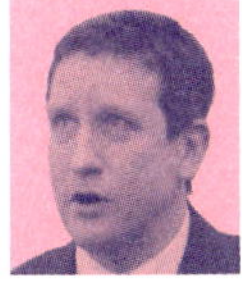

Kazys Varnelis
Netlab
Columbia GSAPP

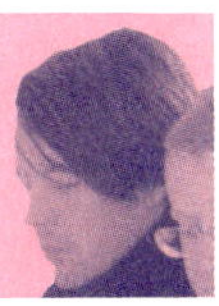

Daniel van der
Velden
Metahaven

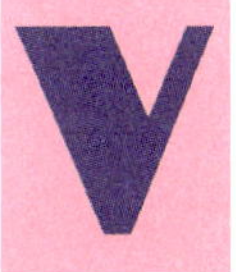

Verso

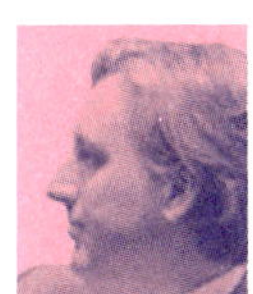

Tim Ventimiglia
Ralph Appelbaum
Associates

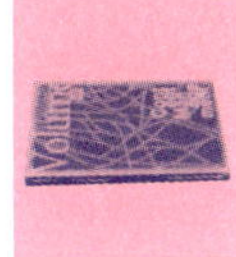

Volume Magazine

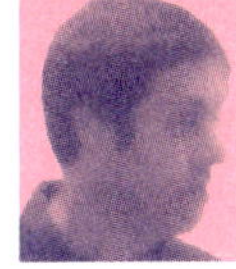

Enrique Walker
Columbia GSAPP

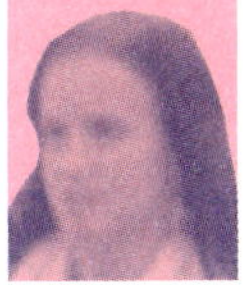

Miriam Walls
NYC Dept of
Education

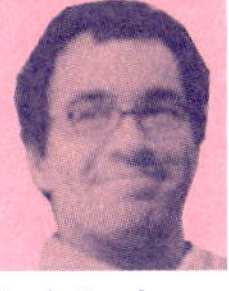

Jamin Brophy-
Warren
Wall Street Journal

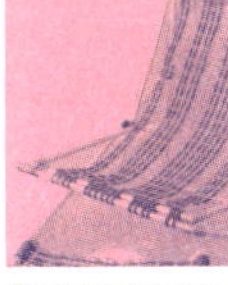

Christine Warren
Redst/Collective

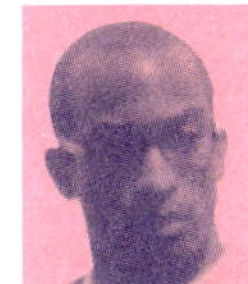

Dorian Warren
Columbia SIPA

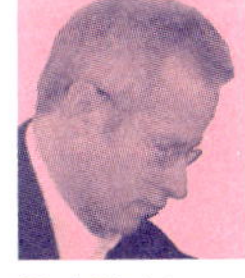

Mark Wasiuta
Columbia GSAPP

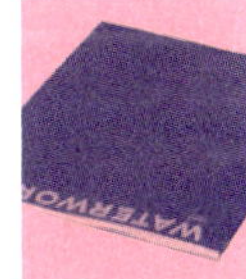

Waterworks

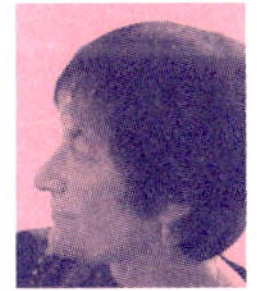

Linda Weintraub
curator

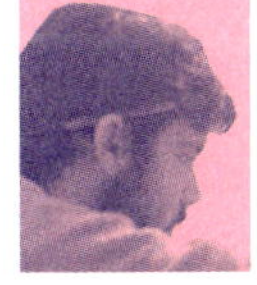

Elliott White
Pratt Institute

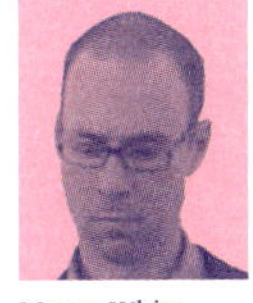

Mason White
University of
Toronto

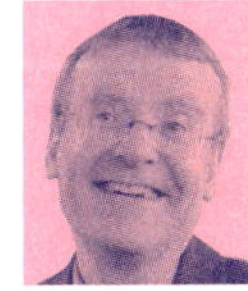

Mark Wigley
Columbia GSAPP

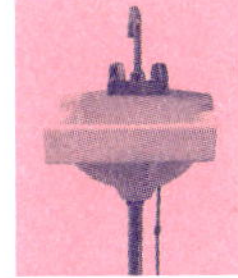

Cathy Wilkerson
writer

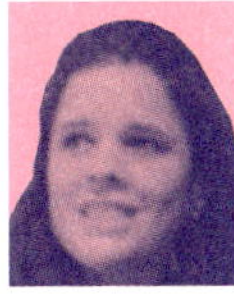

Sarah Williams
SIDL
Columbia GSAPP

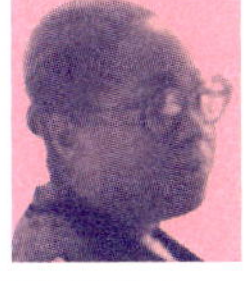

Mabel Wilson
Columbia GSAPP

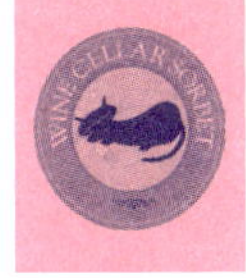

Wine Cellar
Sorbet

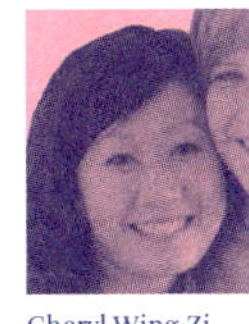

Cheryl Wing-Zi
Wong
Warm Engine

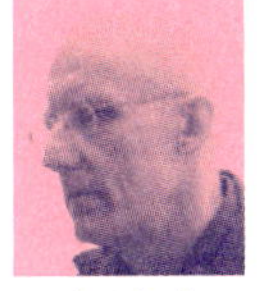

Pawel Wojtasik
filmmaker

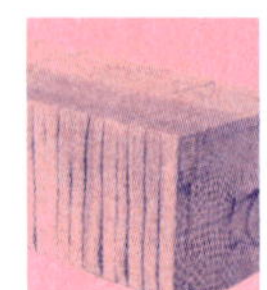

Tobias Wong
designer

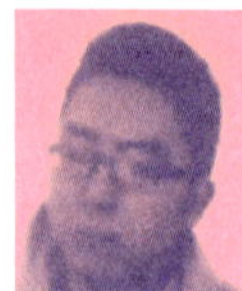

Eric Xu
Abitare China

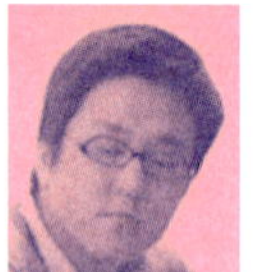

Soo-in Yang
Living Architecture
Lab
Columbia GSAPP

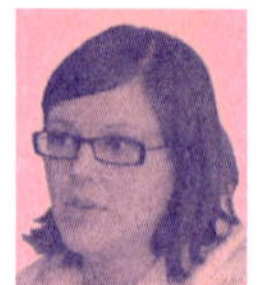

Andrea Zalewski
Columbia GSAPP

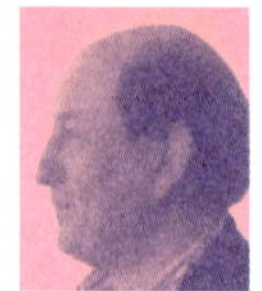

Mirko Zardini
Canadian Centre
for Architecture

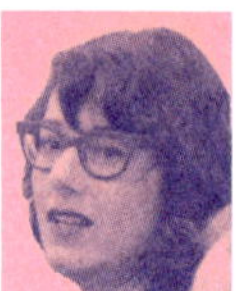

Mimi Zeiger
loud paper

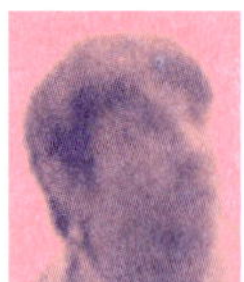

Jason Zuzga
writer

Activate.
Events need not conform to any particular typology. In fact, while some events are easily classified, many defy any fixed definition. The following chronological compendium presents a cross-section of past activities at Studio-X New York, as described by those who helped to activate the space.

<u>Neighborhood Watch:</u>
<u>Joshua Prince-Ramus</u>

August 14, 2008

In the inaugural Studio-X public program, REX President and Studio-X New York neighbor Joshua Prince-Ramus shared his work and design philosophy with an audience of peers: tenants of 180 and 160 Varick Street. Of this image, Prince-Ramus states: "We believe architectural agency can be resurrected if piggybacked upon the owner's needs. Metaphorically, if the owner's constraints form a round hole, why not pick an embodiment of your agenda that is a round peg? Then your vision will slip right through the owner's needs without either being compromised."

Raja Shehadeh
October 16, 2008

On October 2, 2008, writer and lawyer Raja Shehadeh recorded a walk through the hills near his home in Ramallah with his informal walking group *Shat-ha* (Arabic for "picnic"). A video was shot by Shehadeh and Bassam Almohor—neither are professional videographers, and they made no attempt to stage anything. The walk began at 6am at Ayn Arik, then headed over the hills, with two breaks, to arrive at the village of Saffa at 1pm. The footage was sent to Studio-X, where it was screened, accompanied by Shehadeh's live narration via Skype from Palestine (conducted at 1am Palestinian time), fourteen days later.

—

The walk that was captured on video and narrated was not one of the seven walks described in my book *Palestinian Walks: Forays into a Vanishing Landscape*. This time I decided to take a different route, which began just south of the village of A'yn Arik (northwest of Ramallah), heading in a northwesterly direction towards the village of Beit 'Ur el Foqa. The walk began on a dirt road and then moved along paths and alongside several hills, down into the *wadi* and up again. To the south could be seen parts of the Separation Wall that Israel was building on Palestinian land, as well as a new highway that cut through the hills—causing yet further destruction, connecting the illegal Israeli settlements to each other.

Usually I like solitary, silent walks. This had to be different. I had the feeling of being accompanied by a crowd of people whom I have not met, sitting in a studio far away from Palestine, who were dependent on me for a description of what the camera was recording.

For the sake of my sanity, when I walk in the Ramallah hills, I try to avoid seeing the destruction caused to the landscape by Israeli settlements and roads, and the Wall. Instead I concentrate on looking at what remains unscathed. This time, I could not be discriminating. I had to communicate with an audience unfamiliar with this land, and so had to show and describe everything that could be seen: the beautiful, the ugly and the perilously illegal.

—Raja Shehadeh

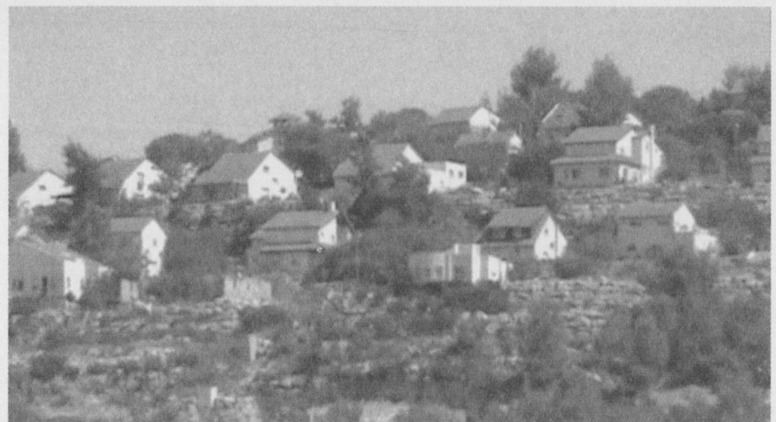

Settlements

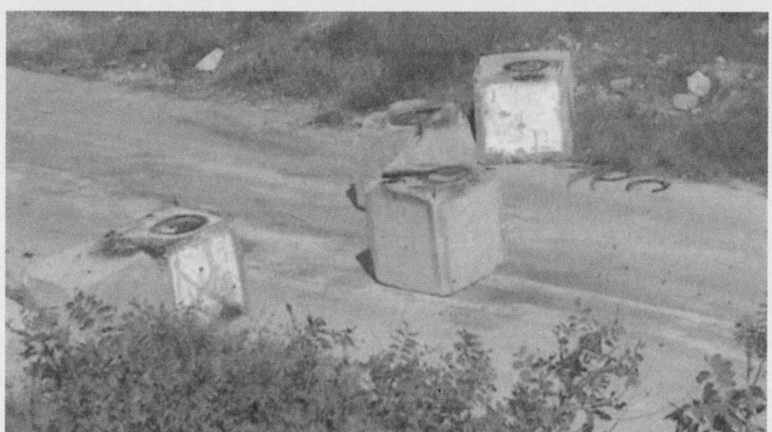

Roadblocks

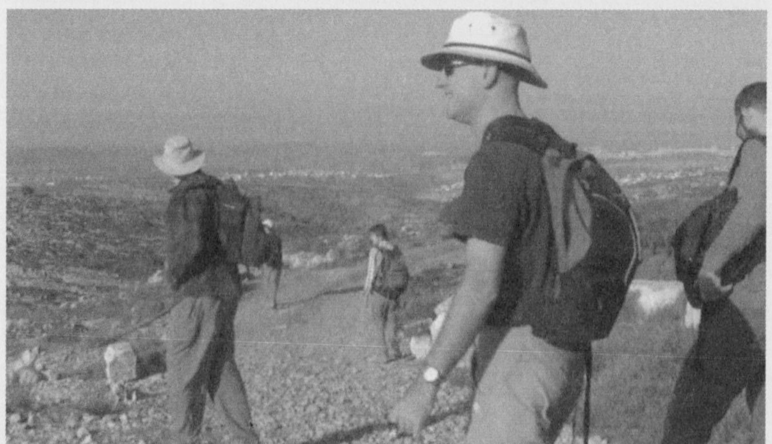

Live from Palestine

Gavin Browning, Cristina Goberna,
Matthias Neumann, Minna Ninova,
Sierra Feldner-Shaw, Abbot Street,
Cathy Wilkerson

October–November 2008
—

"Why did the Weathermen only attack buildings?," I asked Cathy Wilkerson at the final session of the Studio-X reading group centered around her book *Flying Close to the Sun: My Life and Times as a Weatherman.* "It is an interesting question," she answered after listening carefully. "I will think about it."

The reading group, organized by Gavin Browning in the fall of 2008, was everything but what you would expect from a workshop around the 1960s. Instead of revising history, it fostered reflection on the vibrant times before Obama's presidential election; instead of being nostalgic, it was based in a critical approach to the events related in the book; and rather than being centered on the star appearance of the author, it highlighted the question that the writer formulated for us in the introduction to her book. That is, "How do young people make sense of the violence around them?"

In attempting to answer this, I could have written about global violence as opposed to local conflicts, about the media activism carried out mainly by young people, or about the paralyzing effect created by economic responsibilities. However, it was the group's general approach to the topic, and the illuminating presence of Cathy during the last meeting, that should be highlighted. In the end, I understood that the book was an excuse, and that the valuable thing to be drawn from the experience was a call for awareness as opposed to the passivity of the mere observer. It was a reminder of our political capabilities, and an exercise in resurrecting our responsibility to the future.

—Cristina Goberna

How do we make sense of the violence all around us? A cursed, impossible question. But after reading about Cathy Wilkerson's struggle with that question and the way it utterly defined the moral architecture of her youth, I've begun to outline what is sure to be an forever incomplete thesis: we aren't ever supposed to make sense of violence. Our battle, as it were, is to make violence as intolerable and absurd a feature of contemporary society as we can, using whatever talent or compulsion we have. The problem, as Cathy discovers in her years as a young activist, is that violence is a seductive, adaptable concept—it saturates our culture so fully that we often don't know where to start. Violence is shocking and it's familiar, expected and unexpected, institutionalized and outlawed. As a reading group, I know we enjoyed talking about it, even watching it as a packaged aesthetic in the form of Haskell Wexler's film *Medium Cool.*

So where do we draw the line? Do elegant distinctions and nuance play a part? Do we reject all violence or do we use what we've learned from watching the violence to combat its perpetration? Cathy's honesty in describing her own ethical inconsistencies gave me the intellectual breathing room to both trust my own instincts and continue to question their foundations.

One thing the memoir destroyed for
me is the romance of a politically radical lifestyle. But I won't miss it. Like
so many other parts of the book, this
gentle re-education has only brought
me closer to figuring things out for
myself. It also made for one hell of a
pre-election read.

—Minna Ninova

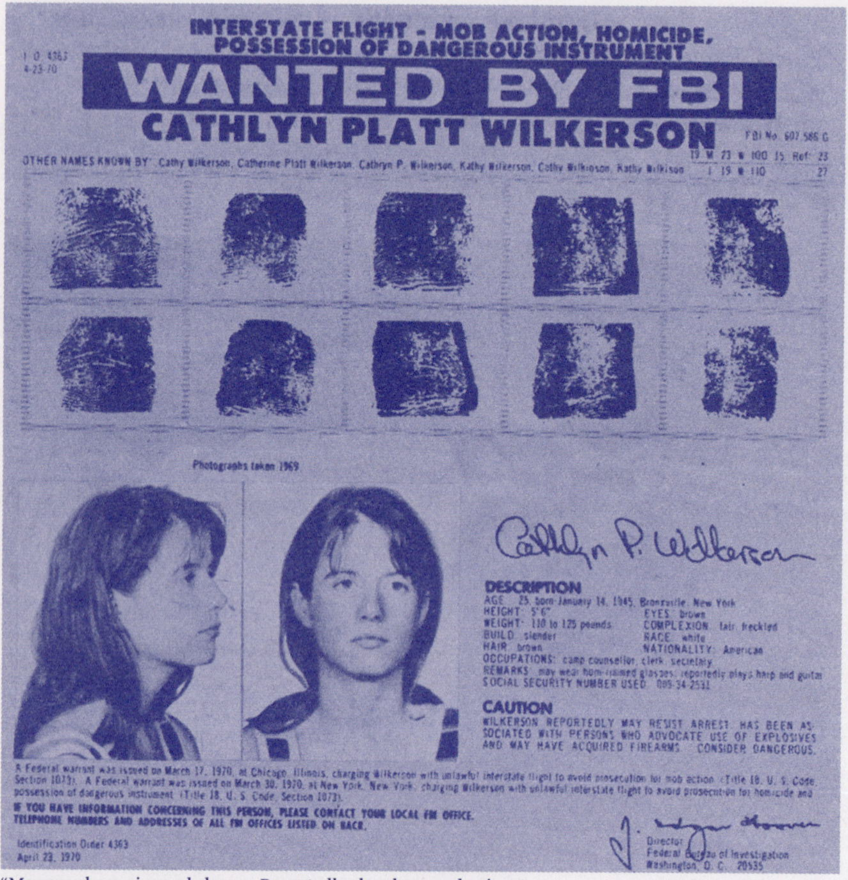

"May wear horn-rimmed glasses. Reportedly plays harp and guitar."

Gavin Browning
I led this reading group around the
election in 2008. The Monday evening
that you came to Studio-X New York
was right after the election. Were you
surprised that a group at an architecture school was reading your book?

Cathy Wilkerson
I'm immersed in the worlds of education and 60s politics, so it doesn't
surprise me that there are pockets
of activity, and that people are making
connections that haven't crossed
my radar. I always assume there's a lot
going on that doesn't cross my radar.
I found it interesting. I didn't find
it shocking. Your explanation of the
space—about trying to be edgy and
to help people make connections—
made a lot of sense to me. You know,
the whole notion of people making
connections, non-traditional connections. That works for me, because
here I am in the education world,
and it was wonderful for me to spend
a night with people in these other
worlds, to hear other people's take on
the book.

GB
I'm glad.

CW
I know there were a couple of people
there who were not originally from
the United States, and I remember finding that very fascinating.
    As far as the architectural part, I
interpreted that through the lens that
you said: that it was about Studio-X
and that you were trying to define the
space and explore the edges of how
you could build a space. Instead of it
being a venue that focused on a certain
menu—a traditional menu of items—

that it would be eclectic, finding an
agenda around connections.

GB
I don't even know if it was so fully formulated in my mind at the time.
That was very early on. Did you receive
the reaction paragraphs?

CW
I don't remember.

GB
They were really interesting. We took
a question that you pose in your
introduction, 'How do young people
today make sense of the violence all
around them?' I don't think anyone had
an easy answer, but we were dealing
with that throughout the process.
One person said that your book made a
really good "pre-election read."

CW
Why is that?

GB
Well, what is your take on why that
would be?

CW
Because I talked a lot in the book
about change, and about the model of
change-from-the-outside versus
change-from-the-inside. The book talks
about starting from the inside, and
abandoning that, and going to the
outside. It talks about ending up being
much more reflective, and feeling like
neither model had a corner on the
market in terms of effectiveness. I'm
still wrestling with that, because
I've been working from the inside in the
world of education, and I've about had
it with that.

GB
How so?

CW
You know, feeling like if there isn't a revolt of the parents soon, then they're never going to figure this out.

GB
The parents?

CW
Children's parents. In the 60s, they shut down the schools because of the hideous racism in New York. Brooklyn had double sessions, and no regular teachers. Manhattan had full-day sessions. It was really…out there.

GB
Education was a big thing we talked about that night. Math, in particular, and the sense of empowerment that you seemed to have gained by learning and teaching math as an adult in the New York City public school system [after you emerged from living underground].

CW
That's still my big push. But I tell you, the problems in the schools in Brooklyn and the Bronx are so enormous—it's really hard to make any headway. I don't teach now. I work with teachers in various schools. They're all schools in various states of disrepair and crisis. So, it's pretty raw.

But, anyway, that continues to be an interesting theme for me. There are inside-spaces and outside-spaces, and there are not a lot of spaces that straddle those two other spaces. It's an interesting dilemma for activists.

I mean, there are spaces that do straddle the line, but they're always under fire—like ACORN [Association

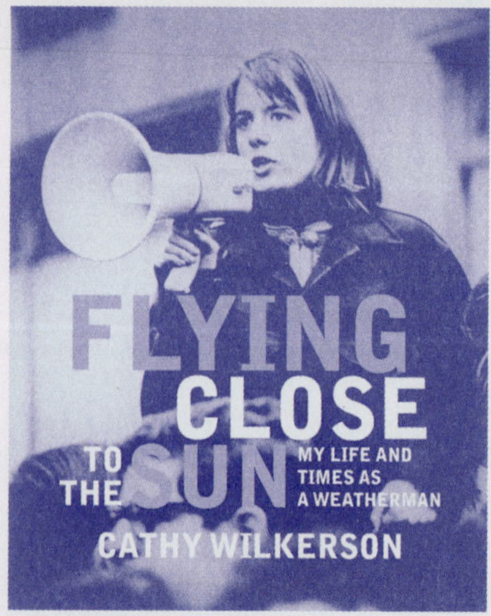

Seven Stories Press, 2007

of Community Organizations for Reform Now]. They're always in-between, and they get beat up for being in-between.

GB
Right, they even recently changed their name.

CW
Yes, well, they screwed up, big time. They didn't get it that they were on the border, and that they had to watch their P's and Q's. But, that doesn't mean that the work they do isn't very good.

GB
That same reaction paragraph from the reading group referred to your story as a "gentle re-education" about the mythos of the outlaw-radical lifestyle associated with groups like the Weather Underground.

CW
Well that works for me! [Laughs] That was one of the major purposes of the book—to put a check on the

*We were looking for a way to inject
fantasy back into architecture—
for the most part, our fantasies.*

—*Felix Burrichter, 01.08.09*

*The hot spot statistic tells us
if the clustering pattern we see has
a neighborhood pattern.*

—*Sarah Williams, 04.07.09*

*I try to be fairly aggressive with
restaurant-menu collecting.*

—*Rebecca Federman, 02.27.10*

*I'd like to know why you chose to do
an exhibition on Renzo Piano.*

—*Sylvia Lavin, 12.05.09*

*I purchased 94 objects of
Leniniana—little busts, sculptures,
posters of Lenin.*

—Yevgeniy Fiks, 12.02.08

*Pressures of urban warfare
are permeating from the war zones
of the periphery of the US empire
into the heartland through
questions of homeland security.*

—Stephen Graham, 04.13.10

*The communities around the temple
[in Belur, India] were fairly orthodox.
They did not want alcohol,
or tourists coming with drugs or meat.
Sustainable Urbanism International
developed a tourism strategy
around vegetarian food,
health, wellness, and yoga.*

— Jyoti Hosagrahar, 03.11.10

romanticization of it all, but not to do it from an equally pompous platform. I was seduced by all that romanticization early on, so I can be completely understanding of people who are.

I think that if people really think about it honestly, they'll realize what it is for itself—and that the world is much more complex.

I still read science fiction, where the good guys are good and the bad guys are bad, and it's really nice to go to sleep to. It's not the real world. It's very comforting to want to go there, but I try to contain that impulse now to science fiction. When you talk to people my age who have any sense, they get more and more humble as they get older. You realize the complexity and the enormity of it all—which doesn't make it any less interesting to be in the fight—but it does change your expectations for success in a dramatic way. I mean your expectations are much more long term.

GB

So instead of some kind of dramatic overhaul of the system that we read about in the book, you're looking for smaller changes?

CW

You realize that those moments of dramatic change are few and far between, and that they happen all the time, but globally. You only get to be in the middle of it every so often. And that also becomes okay. When you're young, you need to have the belief that it can all be fixed quickly. When you get older, you develop a tolerance for imperfection. Not necessarily in your own work, but in the world. I haven't become any more tolerant of imperfection in my own work.

GB

Are you in a reading group?

CW

I think reading groups are great, and I wish I had time to be in them. Someday I'll have time to do that again.

GB

What are you reading now?

CW

A lot of education journals and education books. I literally work 12-hour days, and I have a huge amount of reading to catch up on, and that's what I read on the train. So, enjoy your freedom.

Activate

<u>*re:construction*</u>

Daniel Perlin

November 18, 2008

What do buildings sound like? Artist Daniel Perlin took up the challenge of responding to the Studio-X New York space and program by building a small house—over the course of an hour, before an audience—that contained the sounds of its own making. To make *re:construction*, Perlin used screws, glue, nails, sawhorses, a tape deck, an audio cassette, and a laptop to examine work and construction as auditory processes. The result was a large orchestral work with rhythm, harmony and melody, recorded onto a cassette tape in real time and housed within the new, base-level structure. This performance spawned two further iterations, *re: construction* at MoMA PS1 in December 2009, and *re: destruction* at Issue Project Room in April 2010.

——

Gavin Browning
What is the premise behind *re:construction?*

Daniel Perlin
Urban residents experience the afterglow of construction through its audio production. For example, you rarely see someone actually smashing a wall, but you often hear jackhammers. So you only experience construction visually through difference. You see the initial stage and then the growing stages, and the only relationship between those stages over time is through sound. What evidence do you have of a different equation, besides sound?

GB
Well, you can see changes afoot at Ground Zero, for example. Even from this window at Studio-X.

DP
Yes, when construction itself or reconstruction from nothing becomes the spectacle, it's an amazing and beautiful process. But it's so rarely exposed that way. *re:construction* is a little didactic: its intention is to get people to walk into a building and think, "How was this made? Who made it? What did it sound like?" And now that I'm inside, "How does it sound?" Instead of hearing Muzak when you walked into a building, what if you actually heard the processes of how that building was built?
re:construction at Studio-X was an architectural gesture. For me, it was about reducing the incredible structure of the skyscraper down to its essential procedural elements.

It was also an extension of Robert Morris' *Box with the Sound of its Own Making*—a completely self-contained, sealed box with a cassette recorder within that plays for roughly three hours. My box used a cassette recorder as well, which could play back the duration of the construction, but you could see inside. And it had some of the trappings that a building would offer: protection from the elements, protection from the outside, and an inside.

GB
Along with its interiors, you're exposing the labor that goes into the construction of a building.

DP
Yes, labor is key. In the process of doing the piece, I show how difficult it is to make things. I'm very bad at model-

building and carpentry. During the course of the Studio-X performance, I hit my thumb, and at the second performance at PS1, I drew blood.

GB

Can you describe the process?

DP

In *re:construction*—a building with the sound of its own making—the building builds upon itself. It begins with a hammer, then a drill, and finally a ratchet. Those were the three essential elements, and each of those spiraled off into a digital build, building upon each other and themselves to create a digital musical layer on top of the physical layer.

It's basic sampling. I take a sample from the hammer right there in front of people and I drop it into a series of samplers, and I take a sample from the drill and I drop it into a series of samplers and filters, and with the ratchet, the same thing. The whole piece is built from the sound that I take right then and there.

Walk into a building like this one [laughs, gestures to Studio-X], and you see rows of people sitting in front of computers. What is a building now besides people stationed at digital hubs? The digital meta-layer is a very modern material—a base-level modern material.

GB

So the "re" in *re:construction* refers to this meta-layer?

DP

Yes, it's using digital labor to take noise and transform it into sound.

GB

Why didn't you build a skyscraper, especially for a performance in New York City?

DP

I wanted to try to find the base possible elements necessary to construct a building, not a box. Instead of building four solid walls, I left an opening to imply a door. I was hoping that the roof would get covered, but I misjudged—again, I am a very poor carpenter—and it ended up with a gap. Some people said that the finished structure resembled a favela, and that's fine, because the piece is intended to make the most base-level structure. A favela requires the least expensive materials available, and it's often constructed with limited labor time.

Regarding Studio-X in particular— being on the 16th floor of 180 Varick Street, overlooking all of Manhattan, considering post-*Delirious New York* ideas of the skyscraper as a metaphor —I thought that this space and this view would add an extra layer of complexity. After I finished, I knew I couldn't perform the same *re:construction* anywhere else. It requires this space, and these windows.

<u>Soviet Contamination</u>

Barry Bergdoll, Jean-Louis Cohen, Yevgeniy Fiks, Vitaly Komar. Moderated by Jorge Otero-Pailos.

December 2, 2008

What is (and is not) being done to preserve the architectural heritage of the Soviet era? This panel—celebrating the five-year anniversary of historic preservation journal *Future Anterior*—addressed this question from a multitude of angles. The issue's guest editors, Barry Bergdoll and Jean-Louis Cohen, were joined by celebrated Russian artists Vitaly Komar and Yevgeniy Fiks (both of whose work appears here), as well as the journal's editor, Jorge Otero-Pailos.

——

Jean-Louis Cohen
This issue of *Future Anterior* came out of MoMA's exhibition *Lost Vanguard: Soviet Modernist Architecture 1922–32*, which was based on Richard Pare's photographs of buildings erected in the 1920s and early 30s in the USSR.

The worst example of what has been called "preservation" in Russia—in fact complete reconstruction—can be seen in the Moscow Planetarium designed by Barshch and Siniavsky, which was raised six meters to allow two floors to be added *under* the building. Part of the concrete skeleton has been kept. But basically, the building only exists now as an index, pointing to another object that was the real building, now vanished. The Melnikov House in the Arbat District area of Moscow is another lens through which to see the complexity of the situation. Half of the house is still owned by a grand-daughter of

Melnikov, while the other half is owned by one of the most enlightened Russian oligarchs, Sergey Gordeev, who spends part of his money celebrating the avant-garde and its history. Imagine a house that has two competing and conflicting owners—a house that should be preserved not only as an envelope, but as a living milieu.

Most of these buildings have had a series of different lives. Le Corbusier's Centrosoyuz was built by a Russian architect with episodic supervision by Charlotte Perriand, who would go to Moscow to report back to Le Corbusier. The resulting building can be considered a partial betrayal of the original design. So, what are we discussing? Are we discussing the restoration of Le Corbusier's design? Or are we discussing the restoration of Le Corbusier's design and its Stalinist modification? Or, is it the Khrushchevian remodification? The layering of history is amazing, and part of the problem.

These buildings were built for the transformation of society. The problem is not preserving substance—transforming the Narkomfin into a luxury hotel, Melnikov's House into a museum, workers' clubs into casinos—because this has already happened. The problem, I would say, is the preservation of utopia, the preservation of context, the preservation of the traces, memories and programs of the earliest Soviet Union: a society that was aspiring to a radical change in human history.

Barry Bergdoll
As a recently finished graduate student in Paris, where I had befriended Jean-Louis, I longed for a trip to the Soviet Union. This was going to be the great experience. I couldn't go on a tour—that would be too organized.

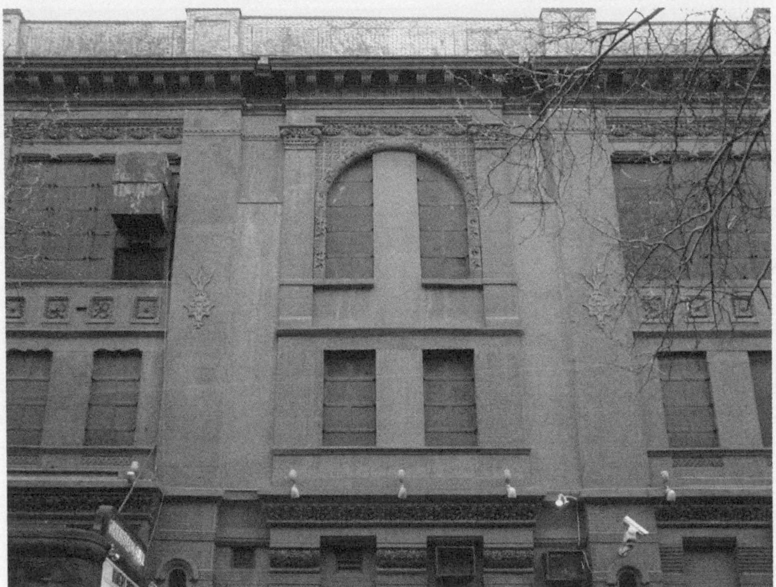

<u>12 East 11th Street. Webster Hall</u>. Earl Browder, a long-time American Communist and a former General Secretary of the Communist Party USA (1932–1945), gave a speech here on March 30, 1950, after his expulsion from the party. He was expelled from the party in 1946 after being attacked by the rest of the international Communist movement for his intentions to distance CPUSA from the Soviet Union. (Photo and caption by Yevgeniy Fiks.)

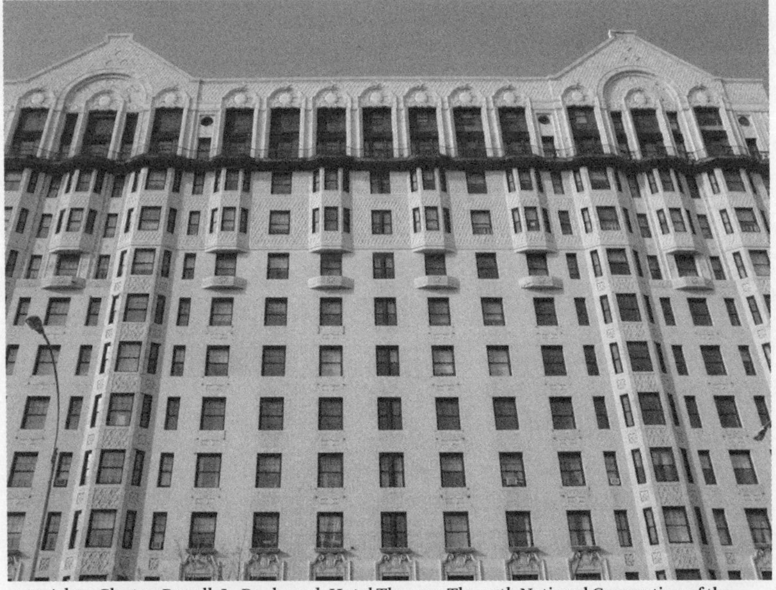

<u>2090 Adam Clayton Powell, Jr. Boulevard. Hotel Theresa</u>. The 17th National Convention of the Communist Party USA took place here on December 10–13, 1959 (as well as many other CPUSA conventions). During party conventions, undercover agents would sit in cars outside the hotel taking photos of everyone coming and going. (Photo and caption by Yevgeniy Fiks.)

I had to go as an individual adventurer. This journey had that old frisson of spy novels. With the invaluable assistance of Catherine Cook's map from *AD* magazine, I was going to find the monuments of Soviet architecture. Even in non-communist countries, these maps had something of a treasure hunt quality; but in this case, the experience was a real revelation and ultimately a shock.

This architecture had been taught to us as so vital to the project of the revolution in the 20s, that when I finally got to Moscow, I expected to find a city that been reshaped by the avant-garde. To discover that, in fact, these buildings rarely had any relationship one to another—in my textbook-trained mind they were family of buildings, in dialogue with one another—and that in urban space they were in fact very far from one another, and indeed that many of them were located in very peripheral areas…that already told me a great deal. One had to walk enormously long distances, much longer than I realized on the map, because I soon learned that the publication of accurate maps was not legal in the Soviet Union. Richard Pare's archive reverses the whole issue. One realizes how thin on the ground the architecture of the 1920s is in Moscow, yet how thick it is on the overall territory—it wasn't just a phenomenon of the capital.

But to cut a long story short, my desire to see the incredible ramps of the Centrosoyuz led to my arrest in the Soviet Union. At the time I was terrified (happily the detention was only a matter of hours, long hours…); but now I truly feel like a child of the Cold War, because I was able to collect the Soviet prison experience, and then in 2007 to hang Pare's photos in The Museum of Modern Art. [Laughs]

Vitaly Komar

I've been in New York since 1978. It's difficult to find an appropriate metaphor to explain Russia's destruction of its past to Americans. Many Russians believe they can change history and make it more beautiful. The best comparison I have found would be a monument from the Civil War in the South. I can't imagine people from New York or anywhere in the North going to the Southern states to destroy these monuments.

Transitional periods of history are very interesting. They remind me of mornings and evenings, or autumn and spring. That's why many people find ruins to be poetic—they are the autumns and sunsets. From my point of view, ruins and unfinished pieces of art are very similar. As you remember from the Renaissance, Michelangelo and Leonardo Da Vinci often did not finish their works. And critics of early modernism viewed painting of the time as being unfinished. We have to appreciate how ruins look.

When speaking about the ruins of the avant-garde, we must remember that the Lenin- and Stalin-era avant-gardes were connected to their totalitarian political systems. The Soviet avant-garde was not innocent. It was much more complicated. We have to understand the negative side of the Russian avant-garde, and positive side of Soviet kitsch.

I've worked collaboratively before— I've worked with Alex Melamid, and I've worked with Andy Warhol. I find the destruction of Soviet armaments to be a very interesting cooperation. In the case of Tomsky's iconic statues of Lenin, people were collaborating with high art: there's a sculpture, and people on the street destroy it. They created ruins.

*We marked out every block
in New York City where there were
three or more foreclosures of
1–4 family homes [in 2008]. Of the
roughly 1,400 markers on the
The Panorama of the City of New York,
only three were in Manhattan.
That gives a sense of how
disproportionate the spread was.
Many of the worst-affected
neighborhoods in this crisis were the
same ones denied capital during
the time of redlining. Now we're
on the other end of it, where the same
people were being given capital
under unfavorable terms. This is what
we mean by predatory lending.*

—Prerana Reddy, 09.17.09

*Of the 536 advertising spaces in
Central Harlem, 135 were for alcohol.
That's 25% of the total ad space.
44% of these 135 ads were located
within 500 feet of a school,
and 24% were within
that distance of a playground.*

—Naa Oyo A. Kwate, 02.27.10

*Silicon Valley has the most
Superfund sites in the US.*

—Amanda McDonald Crowley, 03.25.10

*In New York City,
the hottest spot on the hottest day
[of 2002] was
JFK International Airport.*

Nora Libertun de Duren, 03.05.09

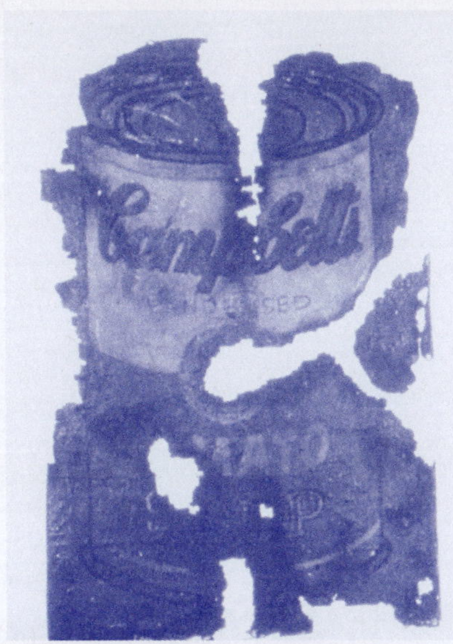

Post-Art #1 (A. Warhol), Komar & Melamid

Therefore, this action is a new cooperation between different artists.

Yevgeniy Fiks
Many of my projects deal with the American Left, and specifically the history of the communist movement in the US. For a project called *Adopt Lenin,* I purchased 91 objects of *Leniniana*— little busts, posters and sculptures of Lenin. Over the last fifteen years, many of these objects have become precious commodities, being sold on eBay, gift shops and antique shops. The idea was to remove those objects from the market and to free them from the process of commodification so that they were no longer fetishes.

This collection was exhibited at Winkleman Gallery in September 2008. The objects were free for the audience to adopt. They couldn't take them during the show, but they could reserve them on a first-come, first-served basis, and then when the show closed, they could pick up their object, free of charge. But in order to do so, they had to sign an adoption agreement, stipulating that they could not ever sell this object—not even to receive a tax break by donating it to an institution. Adopters could pass their object along to someone else, but they would have to make this person sign the same contract, so that every object always has a guardian, and so that the preservation process will continue.

I like Susan Buck-Morss' idea that we are all Soviets—Easterners and Westerners alike—because we lived through the second half of the 20th century together. Another project, *Communist Guide to New York City*, deals with the communist legacy of New York City. It is a collection of walking tours as well as 76 photographs of buildings, public places and sites in New York City that have historical communist significance, and specifically, significance to the Communist Party USA. The guide includes the Hotel Theresa at 2090 Adam Clayton Powell, Jr. Boulevard in Harlem, famously the site of the 17th CPUSA National Convention in December 1959, as well as Webster Hall at 12 East 11th Street in the East Village, where former CPUSA General Secretary Earl Browder gave a speech upon his expulsion from the party. Blumstein's Department store at 230-238 West 125th Street was site of boycotts led in 1942 by African-American politician Benjamin Davis because of discriminatory hiring practice.

By looking directly at architecture, the project questions whether the monuments of the American communist movements should be preserved as part of the legacy its 20th-century revolutionary promise.

Jorge Otero-Pailos
Preservation is always a form of co-authorship—a preservationist doesn't work unless there is something already there—and most of you put forth ideas of collaboration and co-authorship. Jean-Louis, your notion that utopia can only be preserved as an invitation to retain traces of another social reality on modernist buildings suggests a form of collaboration across time. Vitaly, you have discussed else-where your portrait of Hitler that was slashed by a spectator—you chose to not have it repaired, but to leave that as part of the work itself, as a form of co-authorship. You let the slasher finish the artwork. Yevgeniy, you presented the idea of the contract as a form of collaboration between artists and own-ers that evokes a sort of social contract. Could one think of preservation as introducing a degree of unfinishedness into an otherwise complete work, as a way to open it up to other collaborators?

YF
My initial idea was to remove the sub-ject from circulation, and the contract was to insure that it was removed and not functioning in the capitalistic economy. As a result, the object is preserved. But *removing it* from the commodification is the fundamental part of the project.

JLC
This question of creating islands in the market economy is exactly the issue with these buildings in Russia today. In order to preserve them, you have sub-stract them, and then insert them in the pseudo-market of cultural goods, which does not exist, or at least does not exist at this scale. Aside from thousand-and-some *dachas* built in the *dacha*

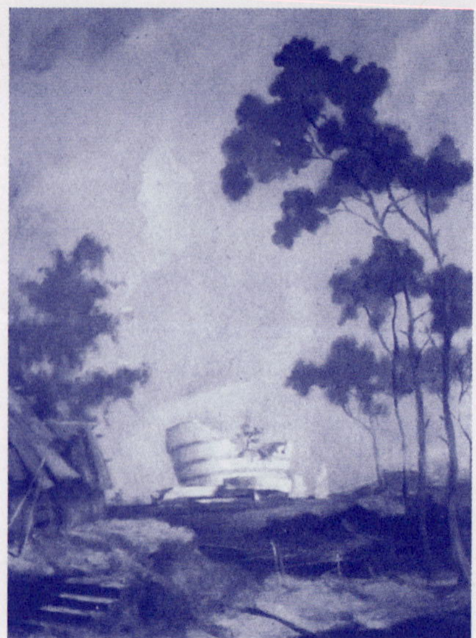

Guggenheim Museum, Komar & Melamid

colonies of the Stalinist elite, there are no houses or small objects in this body of work. We're dealing with big con-tainers that have serious market value today. Therefore, trying to create exceptions—pockets in the system— is very difficult.

JOP
In most Western countries, preserva-tion protects buildings from speculative real estate development, and in that sense puts historic buildings precisely within the sort of pockets of exception that Yevgeniy is talking about. That is in theory at least. We all know in prac-tice the legal shield that protects build-ings works more like a sieve. How are Soviet Modernist buildings protected in Russia today?

JLC
If we want to discuss the current situa-tion in Russia, it's clear that this is an increasingly frozen system, and one in which the free initiatives of citizens are

limited. Government controls almost all the media. The wife of the mayor of Moscow is the main building developer in the city. Political powers and the economy are colluding, and conflict is built into the very notion of trying to preserve some islands of meaning in the booming neocapitalist city.

BB
Of course the only way to preserve the practically unique example of a private house—the Melnikov House—is to make it a public space. The preservation of the object and the preservation of use are in an impossible contradiction. Likewise the way to preserve the Narkomfin, which was an experiment in communal living, might be to convert it into a hotel.

Is this an *exceptionalism* because of the now time-bound history of the Soviet Union—that it has a clear beginning and end? Or are these hyperbolic versions of things that you can actually find close to home? Think of Morningside Heights. The way to preserve the Cathedral of St. John the Divine is to build an apartment house on the southeast corner of the site, whereas the whole nature of the site was to say that it was distinct from the developing apartment house. The only way to do that was to reverse that set of values.

Night Haunts

Sukhdev Sandhu, Andrew Ingkavet

December 4, 2008

In *Night Haunts*, writer Sukhdev Sandhu and composer Andrew Ingkavet presented a sonic journey through an unfamiliar nocturnal London, encountering exorcists, cleaning crews, mini-cab drivers and sleep technicians. For their sound-art performance, the lights were turned out. Sitting in darkness—illuminated only by the surrounding buildings of Soho and the nighttime sky with its stars and passing planes—the audience was invited to face the western windows at Studio-X New York and contemplate another metropolis at night.

—

Night Haunts, commissioned by the rather singular London-based Artangel organization, is a project that involved me wandering and floating through nocturnal London over the course of three years. Wading through subterranean rivers of fat and grease deposited by burger-guzzling and petrol-squandering civilians. Stomping through golf courses on the lookout for urban foxes to shoot through the skulls. Hanging out with exorcists as they tried to rid the city of ancient malignity. Flying above the capital in military helicopters equipped with night-vision cameras that can pick out designer labels from a height of over 2000 feet.

Night Haunts turned out to be, in large part, a journey into the shadowy, occluded world of modern migration. It is, after all, the graft of immigrants, refugees and asylum-seekers that allows the contemporary 24/7 pleasuropolis to

London At Night

June 5, 2008 revisions - C Score

Andrew Ingkavet

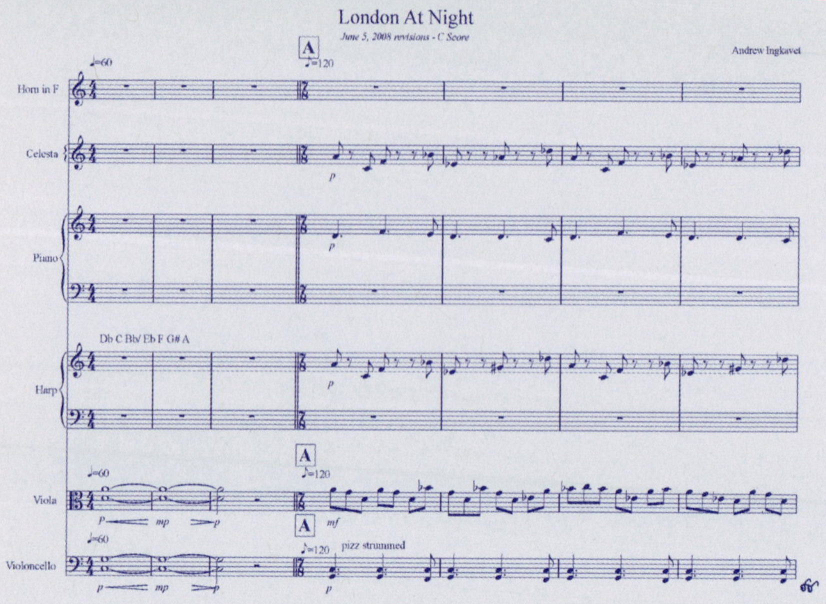

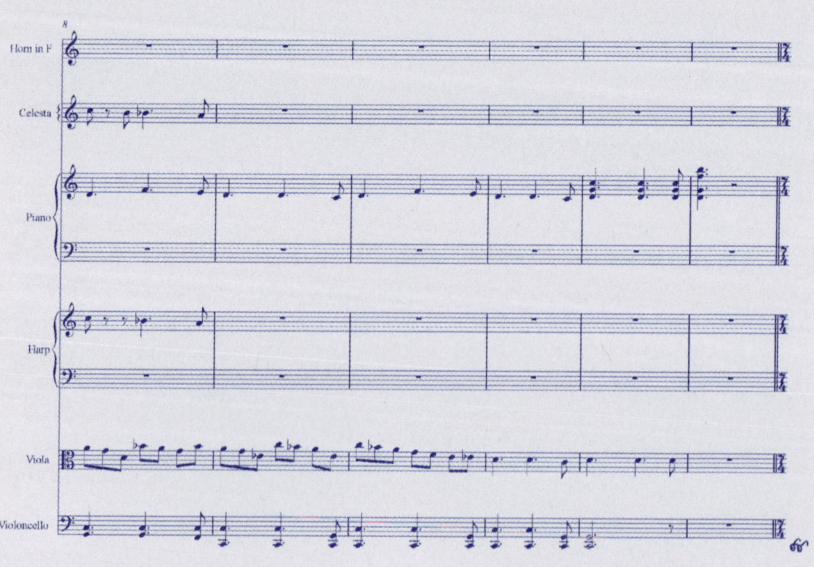

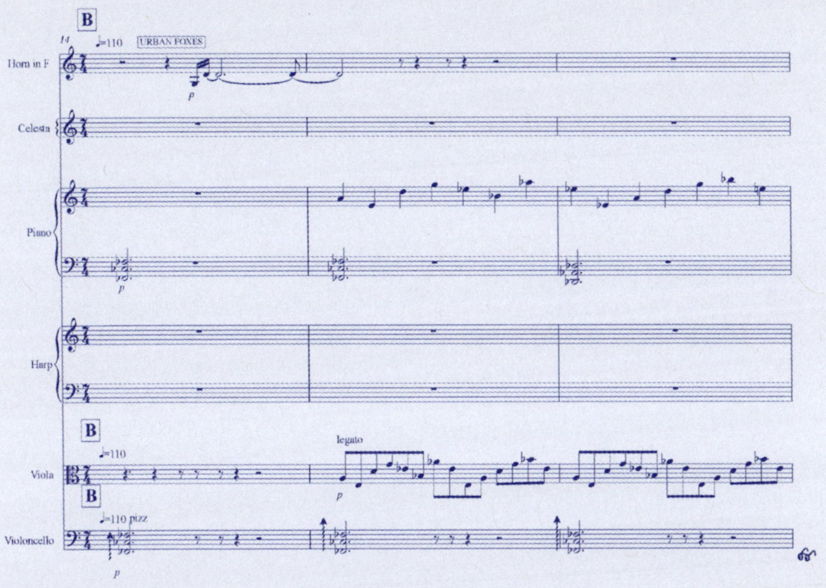

B
URBAN FOXES
Horn in F
Celesta
Piano
Harp
B
legato
Viola
B
Violoncello
pizz

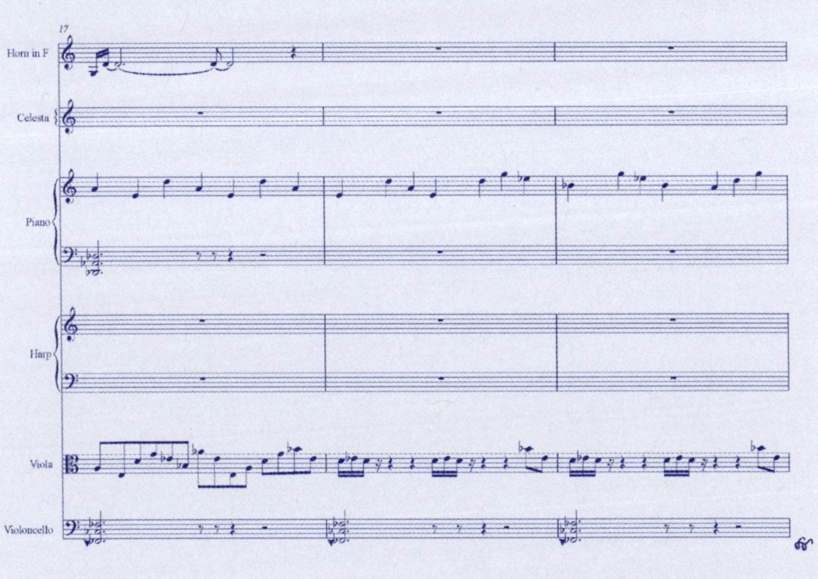

Horn in F
Celesta
Piano
Harp
Viola
Violoncello

flourish. And, though I hadn't expected this at the outset, the end result itself represented a series of migrations: the movement of my own work from its archival and academically-asphyxiating mooring to a place that allowed for looser, imagination-enhanced and speculative writing. Having started as an online project, one in which sound design (specially created by the artist Scanner) and visuals (by Ian Budden) were just as important as its textual elements, various editors, producers and curators were soon asking if it could be reworked into the form of radio essay, television film, and—quaintest of all!—a book.

Gradually I'd become comfortable with the idea that *Night Haunts* was less a discrete entity whose formal contours and parameters could be readily defined than a floating nimbus of obscure topographies and lateral urban analytics that could, in the parlance of dub music, be versioned in many different and equally fulfilling ways. The chance, offered by Studio-X's Gavin Browning, to migrate the project from London to New York, and to translate it into a performance, was altogether too good to turn down: writers always crave, and in today's publishing climate need, to ensure that their books aren't tombstones to creativity, glue-bound artifacts of past graft, but living entities that can mesh and dialogue with other forms of art that reboot and reinvigorate.

I collaborated with Brooklyn-based composer Andrew Ingkavet whose goal, he tells me, was less to "score a book" than to embrace Studio-X's space— particularly its aeriality (different to the catacombish environments in which I'd previously performed) and its vistas (the light-suffused panoramas of Tribeca contrast with often-Stygian London) in order to fashion a "really theatrical event" and "to create an underscore, an emotional webbing that tied the different strands of the project all together."

Neither of us was interested in using the occasion to promote *Night Haunts* as such. We performed in the dark, the better to erase, or at least to obscure, our own presences; we wanted the audience to feel they weren't spectators or listeners, but participants—fellow travelers—in the middle of an off-kilter, intensified auditory environment. Emotion was the key: this wasn't meant to be a lecture, a sideshow, or an argument. Rather, it was an attempt to mesh words and sounds so that, for 45 minutes one December evening in downtown New York, a new kind of space could be felt; a space where the ghosts of contemporary London— its toiler-migrants, unpapered abjects, dreamers—were as palpable and as beautiful as the Manhattan skyline that lay beyond the windows of 180 Varick Street.

—Sukhdev Sandhu

The problem is that nobody is minding the store. Ownership of corporations is extremely diffuse. I own General Motors. I've owned General Motors for many years as a stockholder, and now I own it as a voter. It's very hard to get rid of.

—Moshe Adler, 02.16.10

Much artwork comes out of trauma. We can think of butoh dance coming out of Hiroshima, or Strange Fruit *by Billie Holiday. We can think of* Night of the Living Dead *and the Vietnam War.*

—Karen Finley, 11.10.09

All housing is public housing.

—Reinhold Martin, 06.12.09

*Although synthetic biology
technologies are currently in the labs—
majorly financed and requiring
serious work benches to operate on—
there is good evidence that it
will become more of a DIY field.*

—David Benjamin, 05.04.10

*One role for the architect
and planner is not to build a new
'open city,' but to try
to increase the frequency of openness
in the everyday landscapes
that most Americans call home.*

—Georgeen Theodore, 02.16.10

*The commons are not just
natural resources.*

—Gavin Browning, 03.05.09

Curated by Mimi Zeiger
Opening night panel: Felix Burrichter, Luke Bulman, Stephen Duncombe, Mark Shepard, Mimi Zeiger. Moderated by Kazys Varnelis.

January 8–February 28, 2009

"In the 1990s, zines such as _Lackluster, Infiltration, loud paper, Dodge City Journal_ and _Monorail_ subverted traditional trade and academic architecture magazine trends by crossing the built environment with art, music, politics and pop culture—and by deliberately retaining and cultivating an underground presence. Much has been made of that decade's zine phenomenon—inspiring academic studies, international conferences and DIY workshops—yet little attention has been paid to architecture zine culture specifically, or its resonance within architectural publishing today."

So began _A Few Zines: Dispatches from the Edge of Architectural Production,_ the first exhibition at Studio-X New York. A broadsheet was produced for the exhibition. The oversized document, like the show itself, is a bit unwieldy and insists on taking up too much space, on annexing territory. And like these people on a train, _A Few Zines_ remains in motion. Produced for $150 by _loud paper_'s Mimi Zeiger, _A Few Zines_ has appeared since at Boston's pinkcomma gallery, the Los Angeles Forum for Architecture and Urban Design, and the University of Illinois at Chicago.

—

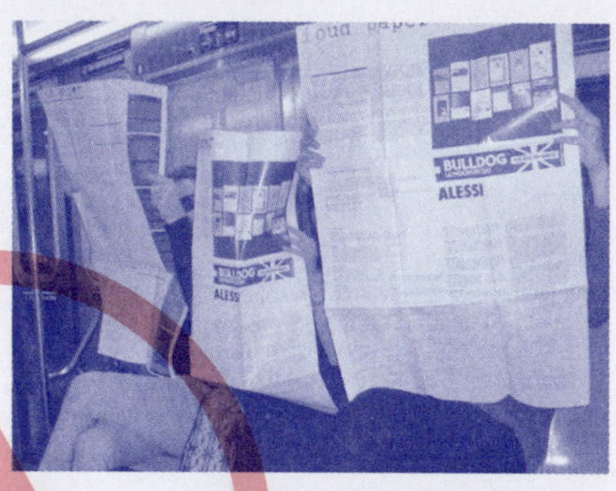

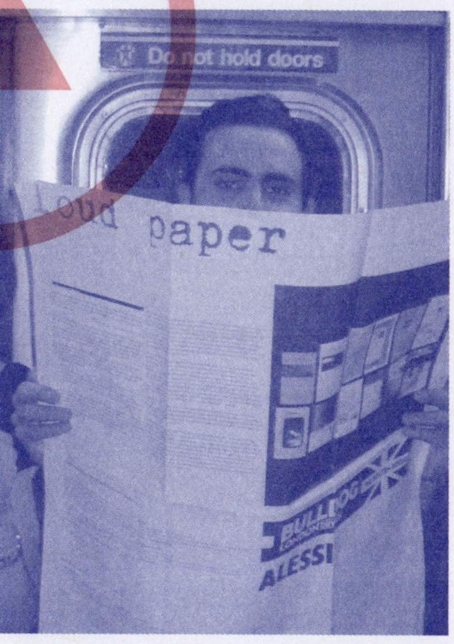

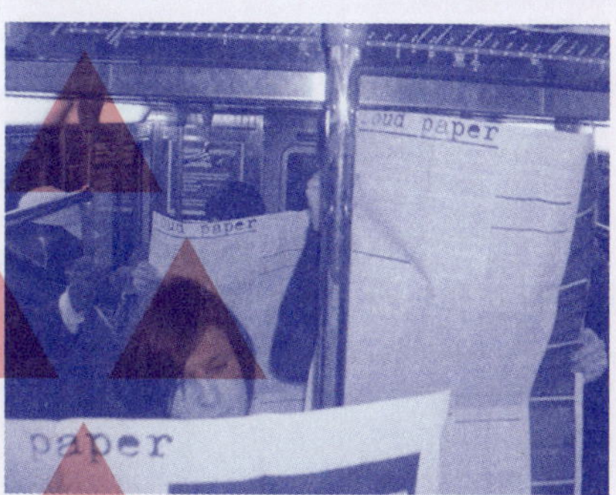

Reading Room

Marisa Jahn, Reinhold Martin, Andrew Ross, Dorian Warren. Moderated by Mabel Wilson.

January 27, 2009

—

On a blistering Tuesday morning in January 2009, I stood bundled, shoulder-to-shoulder with my fellow Americans on the frozen tundra of the National Mall to hear the inaugural address of President Barack Obama. By calling upon us to perform our roles as citizens, President Obama's speech was aimed, as all addresses are, directly at me. The nation's public sphere is indeed vast and diverse, as the 2008 election season revealed. But more importantly, it serves as both a vital space of political engagement and a realm through which we live our public lives. As with any public sphere, we must in some fashion occupy—either actually or virtually—a public space, and that realm must be imagined, built, and maintained. These are the tasks allotted to architects, planners, designers, artists, activists, and developers. Taking cues from President Obama's historic speech, *Rapid Response: Addressing the Address* speculated upon what the quality of that public (and perhaps also private) domain might be.

The evening began with a screening of the address. Then, the panel of educators and activists critically analyzed the message, helping us understand the realistic promises, greater implications and unfortunate exemptions from this speech heard round the world.

They considered these questions in particular:

How will the administration's agenda impact the world that architects and planners imagine, organize and build? How will it address the environmental concerns of sprawl and new energy needs?

How can new national monuments that are planned for Washington DC and New York City address an increasingly transnational culture?

How will Americans address the 21st-century racial politics of space: the de facto segregation of cities and schools, and the hostilities at our southern border with Mexico?

How will his new agenda address the spectacular fall of the housing market by creating new programs for homeownership and affordable housing?

What and whom did President Obama's inaugural address address?

—Mabel Wilson

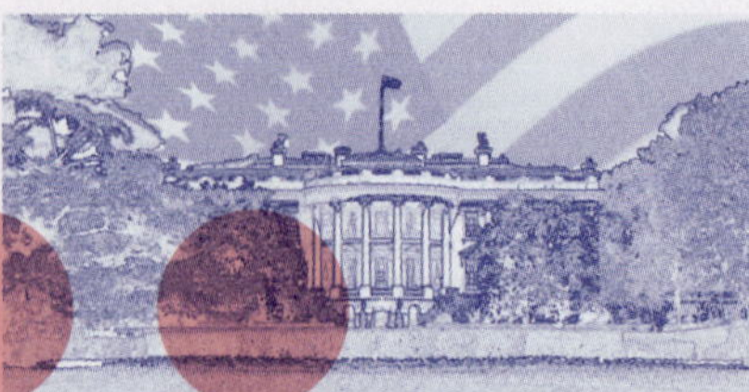

Addressed at Studio-X New York

<u>Book Launch: *J. MAYER H.*</u>

Jürgen Mayer H., Jeffrey Inaba, Andres Lepik, Cristina Steingräber

January 30, 2009

The number of RSVPs for the book launch for *J. MAYER H.* at Studio-X New York far outstripped its seating capacity. The solution? Remove the chairs altogether and request abbreviated presentations from the panelists, thereby fast forwarding to the party—which is what most people seemed to have come out for on a Friday night, anyway. What lacked initially in seating was made up for in the food, drink and conversation later on: music played, people mingled, books were signed and sold, and enterprising prospective authors hovered around the book's editor who, just in from Berlin, was jetlagged.

—

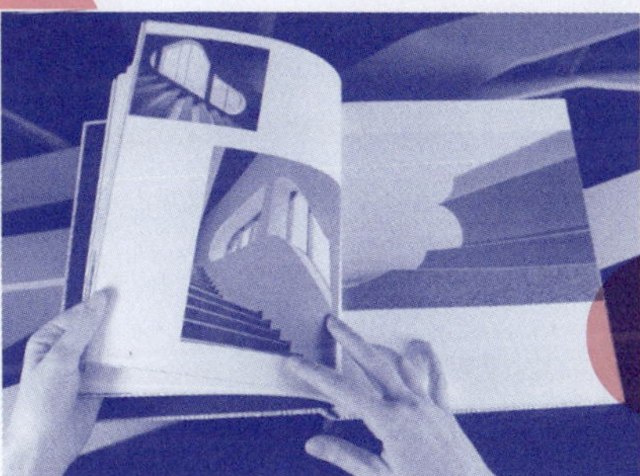

Launched at Studio-X New York

<u>EVERY DAY THE URGE GROWS STRONGER TO GET HOLD OF AN OBJECT AT VERY CLOSE RANGE BY WAY OF ITS LIKENESS, ITS REPRODUCTION</u>

Lars Fischer, David Reinfurt

February 17, 2009

Dexter Sinister and common room came to Studio-X New York to discuss a short-run "library" book composed of texts previously published online as Portable Document Formats and commissioned by the Art Libraries Society and MoMA for a conference on contemporary artists' books. The evening featured spoken-word interruptions from *The Work of Art in the Age of Mechanical Reproduction* and creative use of a circa 1980s overhead projector. When asked to submit something that captured the spirit of their performative presentation, Fischer and Reinfurt sent this text: *ONCE ITS TYPED ITS PUBLISHED* [sic].

—

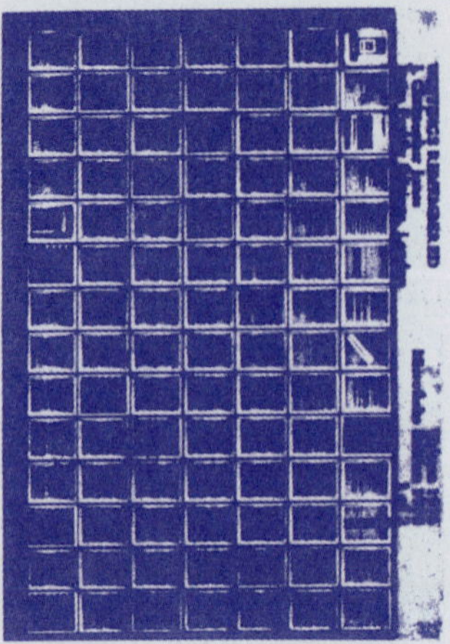

Allen Fisher has asked me to supply a list of my recent writings and to describe any attempts to publish them using new media such as microfiche and xerox.

[photoprint of a microfiche transparency]

The 98 pages of this microfiche, when enlarged to readable size via a suitable gadget (a fiche reader), comprise half of a book which I wrote to explore my records and memories of twenty-five years of writing. The complete book, in microform, can be manufactured and mailed to any place in the world by the fastest airmails for between 5Op & 1 pound (between 1 & 2 US dollars). And this can be done within a week or so of completing the typescript. Its typed 'camera-ready'.

My alternative form of publication is to make xerox copies of the typescript to order, to get them bound by a bookbinder or by a cheaper plastic binding, and to mail them, air or surface, to the persons placing the order. Each book made in this way can be a unique selection of pages from my two thousand-or-so pages of writings: the reader is invited to select the material, and the book is made, and inscribed with the readers name, as an edition of one. The cost of making and mailing a book by this means is 20 to 60 pounds (40 to 120 US dollars). Again it can be done within weeks of completing the text and avoids many of the discouraging difficulties and delays of publishing a book via a printing press.

In proceeding in this way I am ignoring the obvious objections (few people have the gadget to read fiches the cost of xerox facsimile copies is far too high for most readers). These new ways of publishing do not preclude doing so by

ONCE ITS TYPED ITS PUBLISHED [sic]

*In the 1950s, there were 500,000
production jobs in New York City.
By the year 2000, there were
54,000 production jobs
in New York City. The Fashion Center
Business Improvement District
started tracking jobs
[in the Garment District] in 1995.
At that time there were about
16,000 production jobs.
Now there are about 9,000.*

—Barbara Randall, 11.17.09

*Ink intimates hidden relations
between concrete things
and embodied gestures.*

—Michelle Fornabai, 01.28.10

*A citizen-scientist is someone
who doesn't just take in information
from an authoritative source,
but produces it through
the process of exploration.*

—Janette Kim, 10.28.09

*There are inside-spaces and
outside-spaces. It's an interesting
dilemma for activists.
I mean, there are examples that do
straddle the two, like ACORN,
but they're always under fire.
They're always in-between, and they
get beat up for being in-between.*

—Cathy Wilkerson, 03.18.10

*When people start buying protein—
cheese and meat and fish—
that's an indicator of stability
for greenmarkets.*

—Makalé Faber Cullen, 02.27.10

*[With Twitter] you have the
queen bee—Shaq or Lindsay Lohan—
and they are able
to create buzz in places.*

—Jamin Brophy-Warren, 04.07.09

February 19, 2009

What has the Varick Street design community been publishing? ARO, ARROJO Studio, C-Lab, *Metrosource Magazine,* Network Architecture Lab, Phaidon Press, REX, Waterworks and other Studio-X New York neighbors loaned their recently printed and hot-off-the-press works for a one-day exhibition. Promotional materials, books, magazines, articles, take-out menus, catalogs and flyers were all welcomed— as long as they had been published within the past year.

—

Nora Libertun de Duren, Olympia Kazi, Mark Shepard, Michael Mandiberg, Brooke Singer. Moderated by Gavin Browning.

March 5, 2009

Co-sponsored by Eyebeam Art + Technology Center

How can the commons be re-evaluated as both an opportunity and a dilemma today? What implications does this concept hold for structures of sharing, from the circulation of information to the design of urban and public spaces? The diverse work of these panelists, from the fields of art, policy and design, is represented in the following image collections.

—

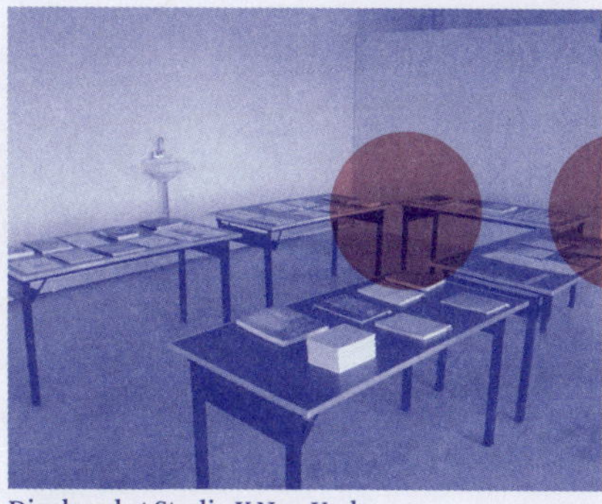

Displayed at Studio-X New York

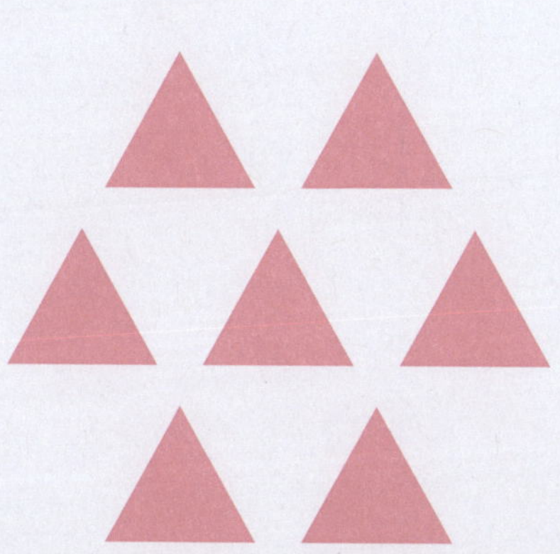

Moderator Gavin Browning screened an animated short to start the conversation. Titled *The Commons,* this film was produced by Browning in collaboration with Laura Hanna, Molly Schwartz and Dana Schechter.

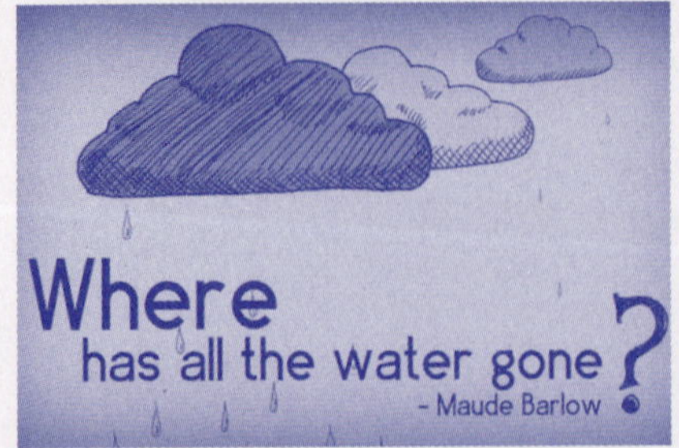

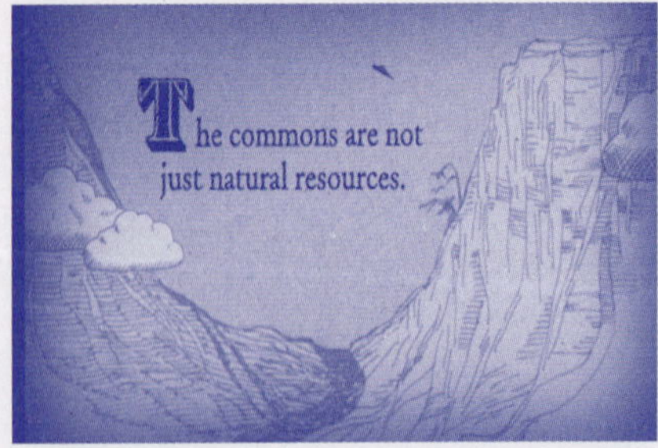

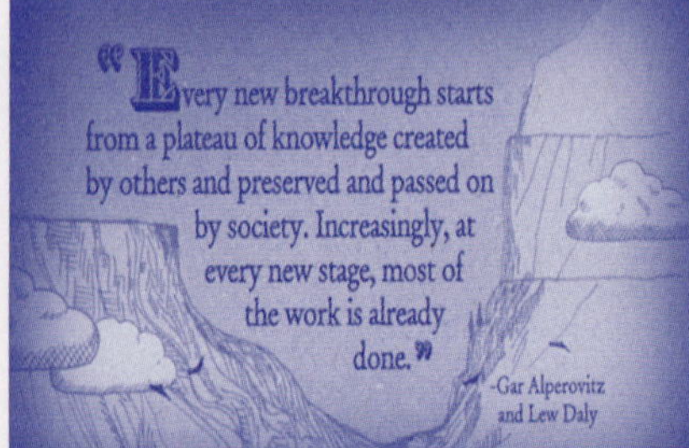

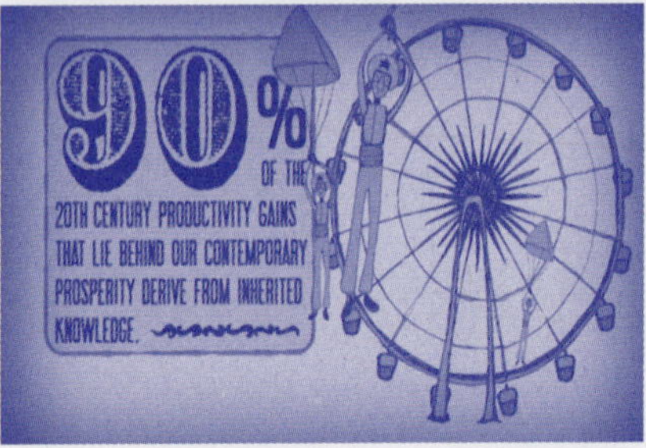

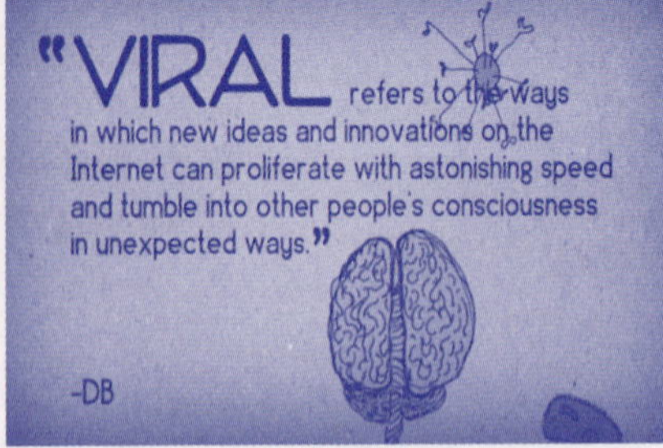

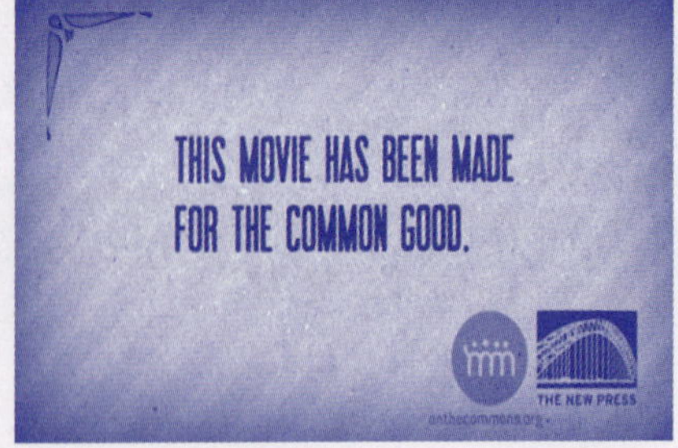

NYC Department of Parks and Recreation Director of Planning Nora Libertun de Duren's vegetation maps help determine where parks should be created across the five boroughs.

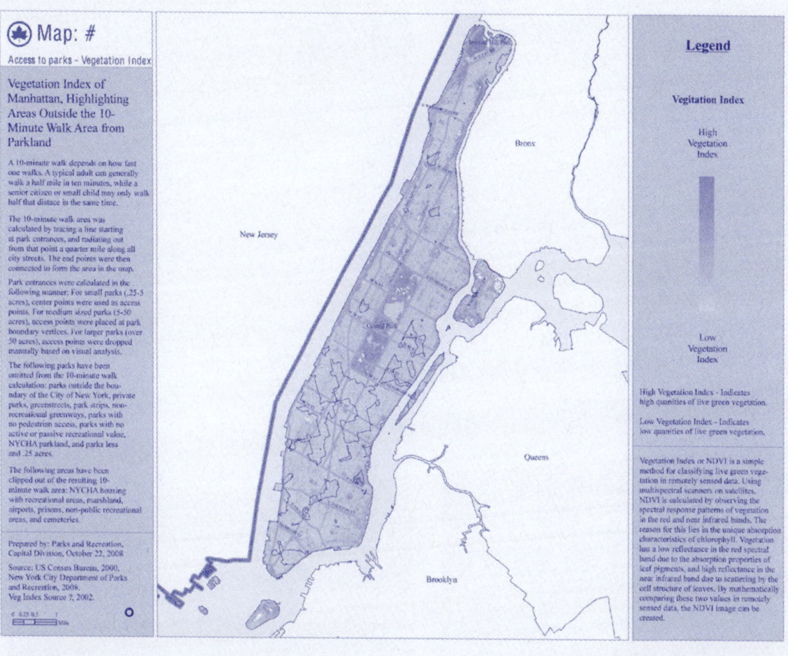

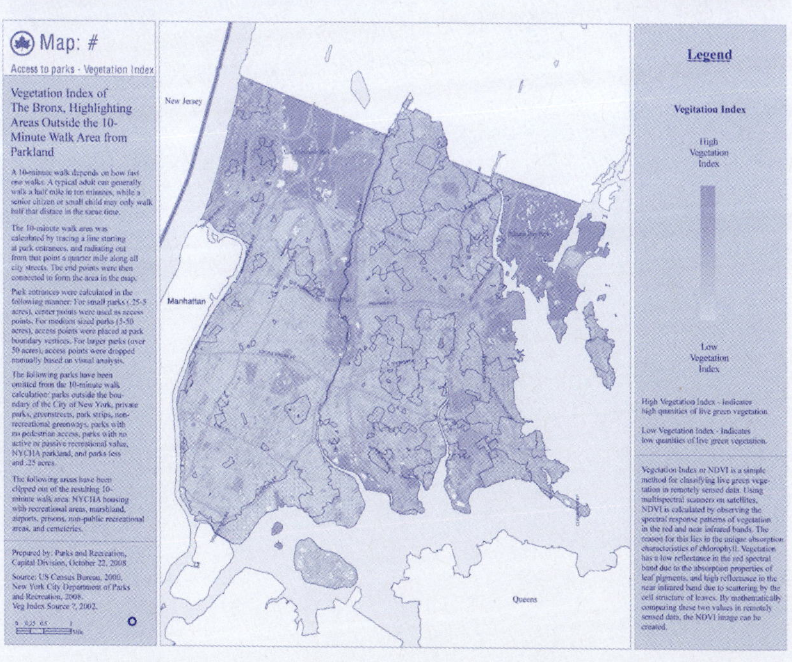

Van Alen Institute Executive Director Olympia Kazi's collection of images show how hard it is to relax in New York City.

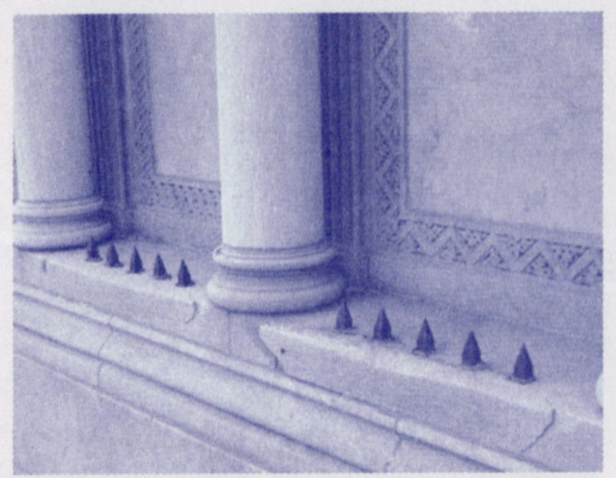

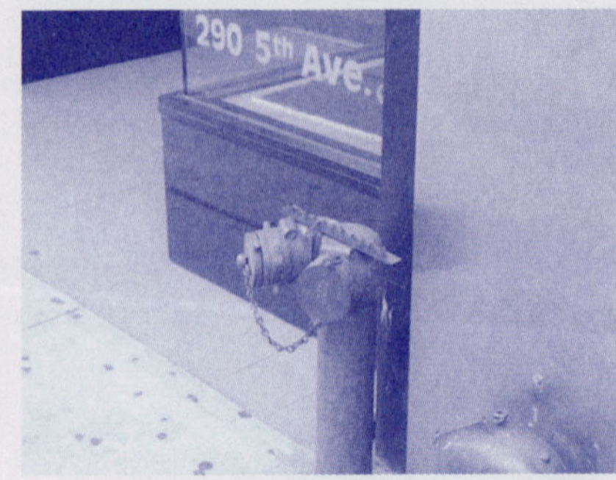

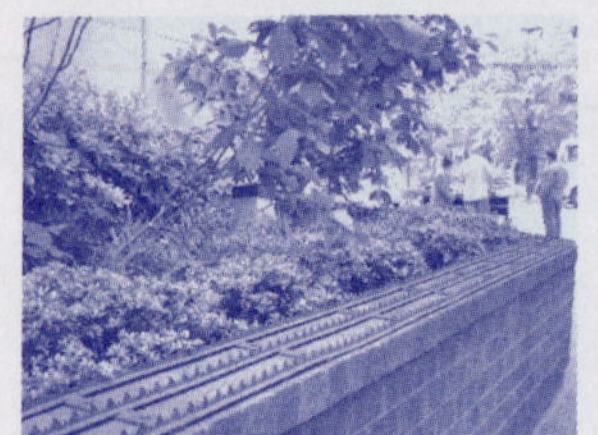

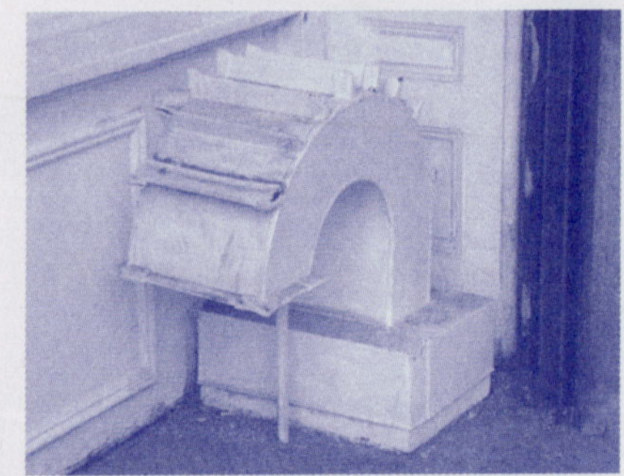

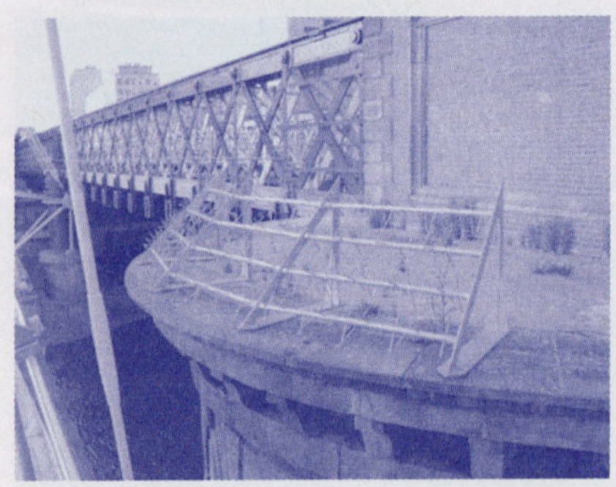

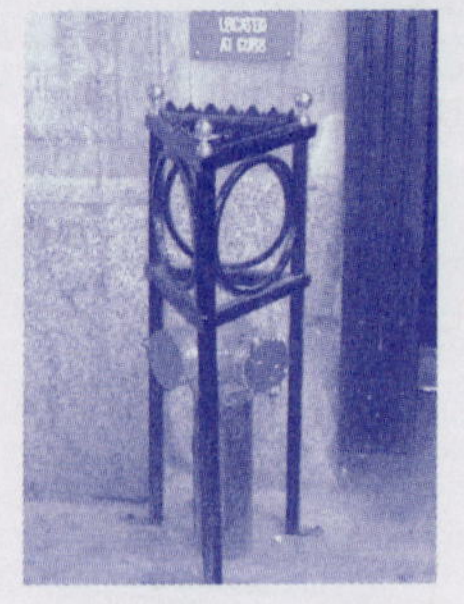

Artist and architect Mark Shepard's *Hertzian Rain,* performed the following evening at Eyebeam Art + Technology Center, explores the electro-magnetic commons.

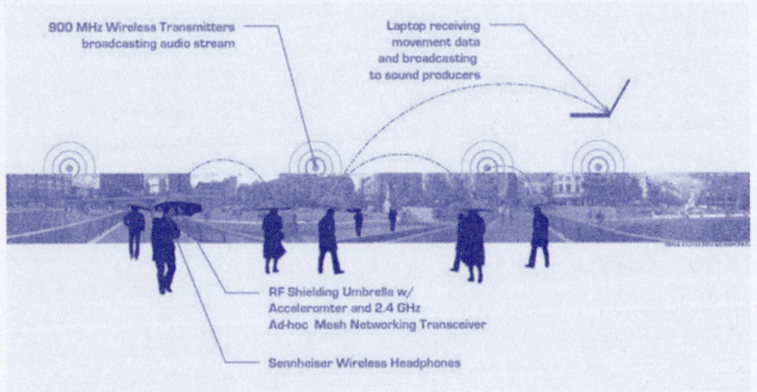

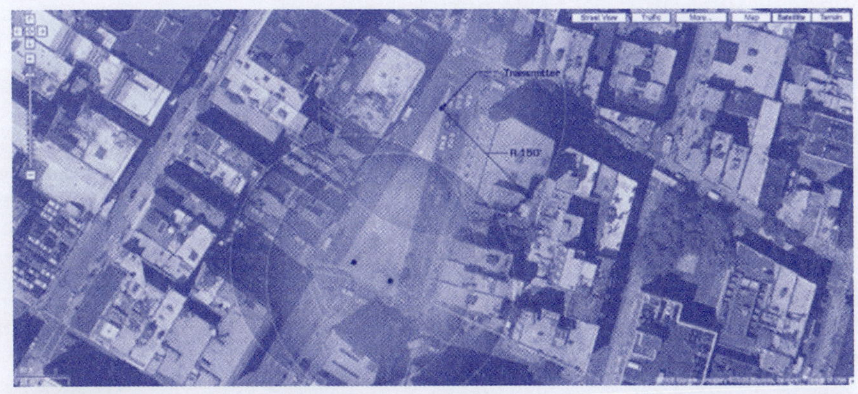

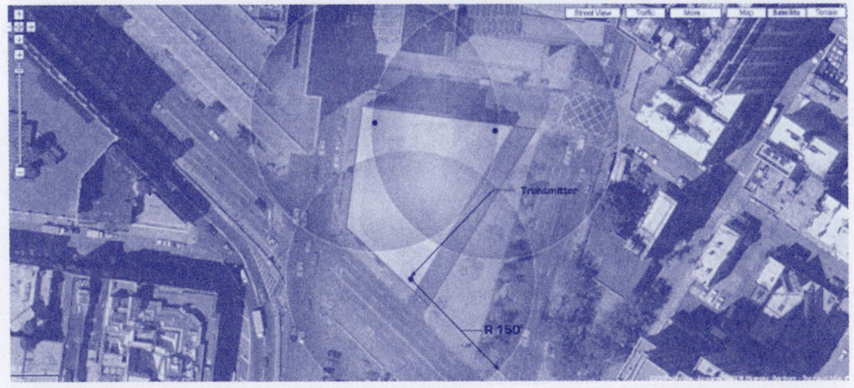

Artist Brooke Singer's photographs of Superfund sites in the Northeastern United States complement her data visualization project *Superfund365*.

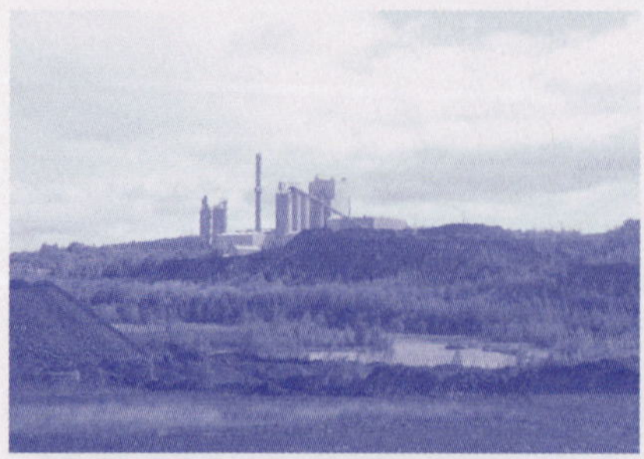

McAdoo Associates, McAdoo Borough, PA

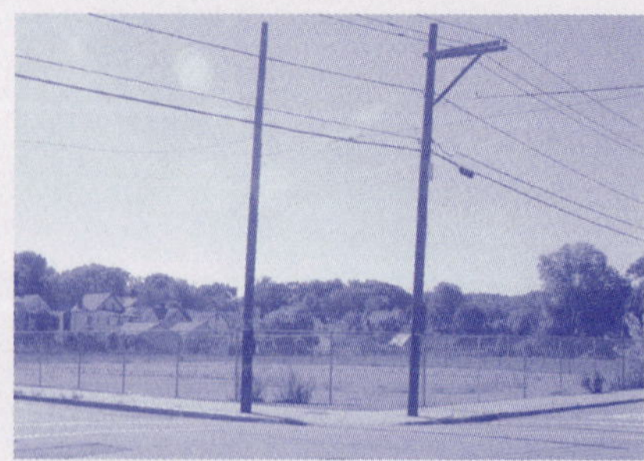

US Radium Corporation, Orange, NJ

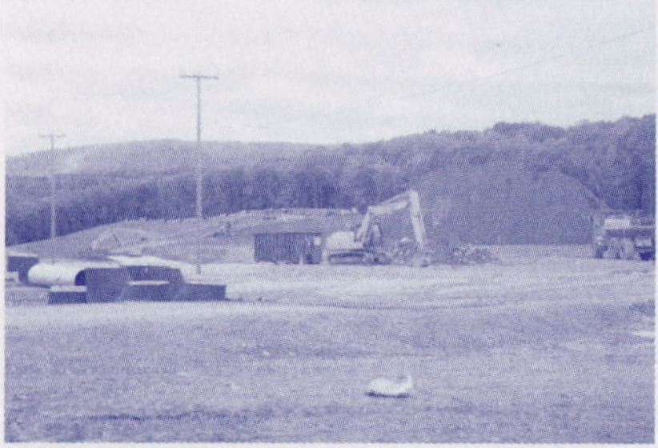

Jackson Steel, Mineola, NY

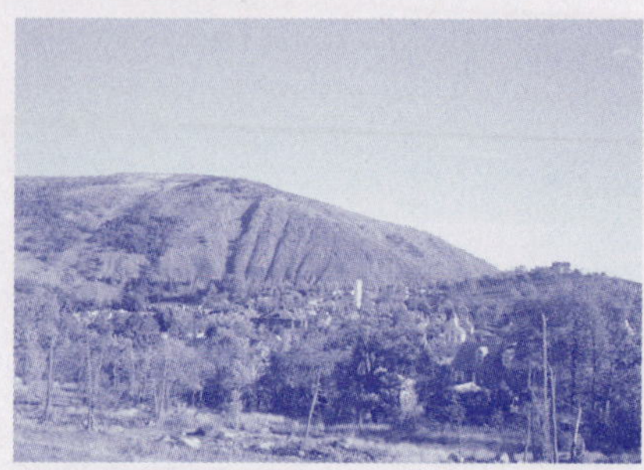

Palmerton Zinc Pile, Palmerton, PA

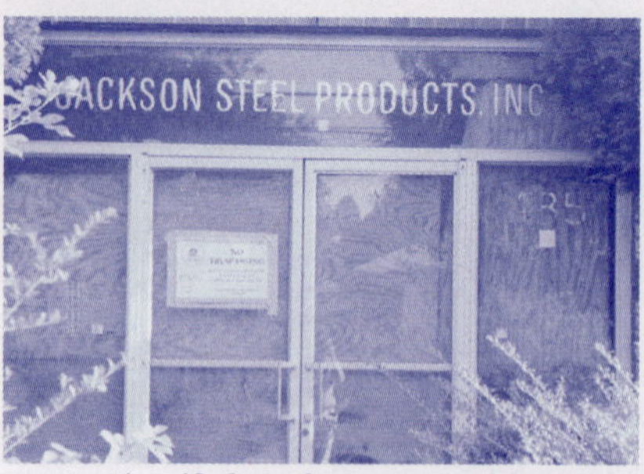

Eastern Diversified Metals, Hometown, PA

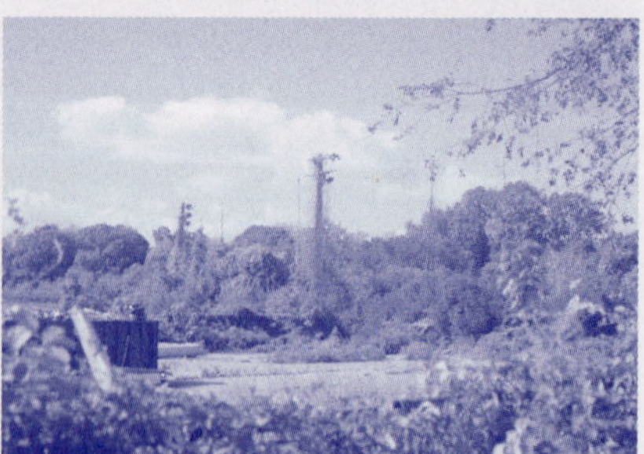

Raybestos Field, Stratford, CT

Artist Michael Mandiberg's photographs of greenwashed garments highlight the phenomenon of pop environmentalism.

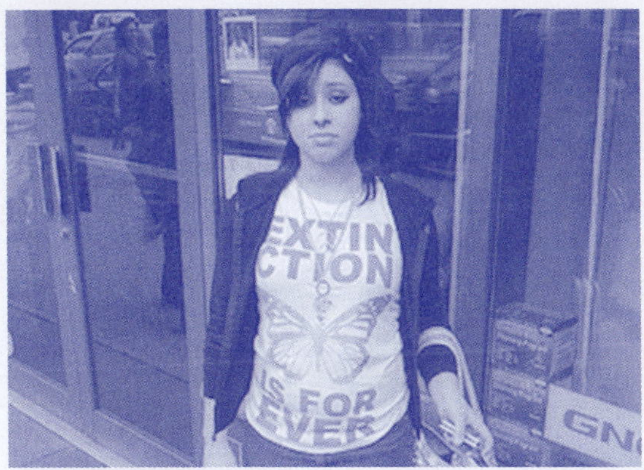

Extinction is Forever

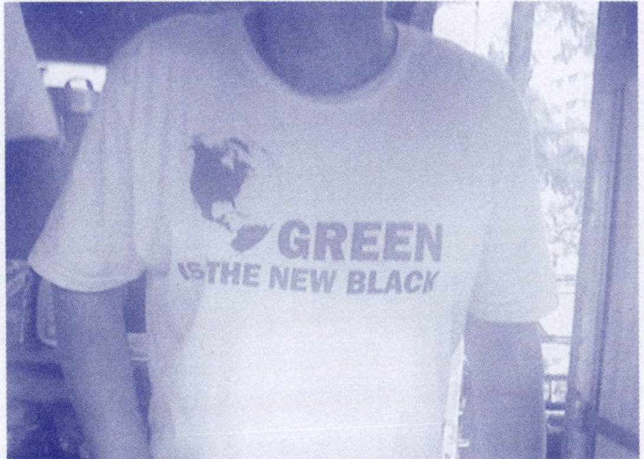

Green is the New Black

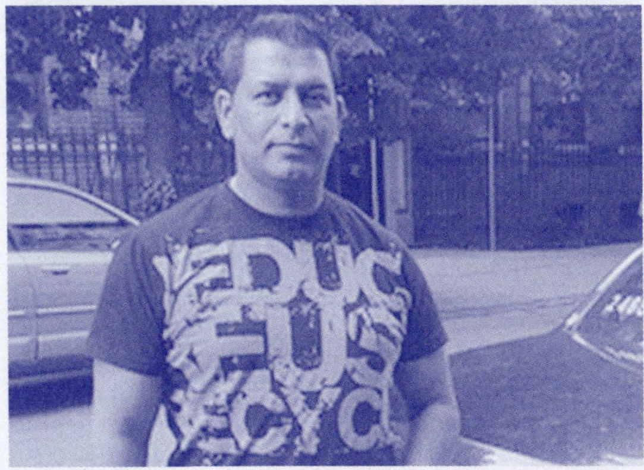

Reduce Reuse Recycle

<u>The Geography of Buzz: Visualizing
Cultural Space in New York and
Los Angeles</u>

**Elizabeth Currid, Sarah Williams
Opening night panel: Erin Aigner,
Elizabeth Currid, Harvey Molotch,
Jamin Brophy-Warren, Sarah Williams**

April 7–May 12, 2009

**This study and accompanying exhibi-
tion by Elizabeth Currid and Sarah
Williams generated nearly as much
buzz as it quantified, landing in
the pages of the *New York Times*, *Time*
and the *Christian Science Monitor*.
What made it so media-ready? Perhaps
it was the striking, large-scale maps
that wallpapered the Studio-X New York
entryway. Or perhaps it was the notion
of the media functioning as an un-
witting urban developer. Most certainly,
it was the caliber of the study and the
design, which drew attention to urban
"event enclaves" by mapping one
year's worth of Getty photographs per
cultural industry in New York and
Los Angeles.**

——

Elizabeth Currid

Parties and social scenes are important
to the cultural industries. I'm sure many
of you agree that the really meaningful
stuff of work often happens outside
of 9 to 5, at times you may not consider
important for anything besides drink-
ing cocktails. Because we know the importance of social scenes, we're limited.
A lot of researchers, myself included,
have tried to get at this—finding that it's
tough as one person. You can study
one art scene in one place; you can go
to three events a night. Well, Getty
Images really opened this up, because

not only do we have access to an enormous amount of events, we also have
an enormous amount of people doing
the job for us. Getty photographers
go to lots of different events at the same
time and at different places. And they
catalog pictures of all of these events.
This offers a very dynamic record of how
creative industries work.

Sarah Williams

One of the great things about the Getty
Images database is that whenever a
photographer shoots an event, they log
where the event took place. This is
where I as a spatial information designer step into the picture. The Spatial
Information Design Lab took each event
in the Getty Images database for the
year 2006, and part of 2007, and categorized the events.

We categorized them into art, fashion, film, music, television, theater
and "magnet," that is, the type of event
that involves a lot of different cultural
industries, but doesn't have one specific
aspect. This might be Magic Johnson's
birthday, for example, or Lindsay
Lohan's New Year's Eve party: magnet,
star-attracting events that don't necessarily relate to any one cultural industry.

We took each event and geocoded
them into a database. Then we performed two types of spatial analysis: the
Global Moran's I and the Getis-Ord analysis. The Moran's I statistic or cluster
analysis allows us to see if the patterns
that we're seeing geographically could
have happened by random chance. If
it wasn't random, we went on to perform
the Getis-Ord Gi* statistic, sometimes
called the "hot spot statistic." This tells
us if the clustering pattern we see
has any neighborhood patterns. So,
while there may be a lot of events
around there, is it significant that it's

particularly high? The Getis-Ord also tells if it's significant that it's particularly low. The results of our spatial analyses produced the maps we're exhibiting here.

The magnet maps reveal one of the big findings from the study—that cities have event enclaves. For example, in Los Angeles, we see Beverly Hills, Sunset Boulevard and Hollywood & Hyland as the key event locations. Los Angeles follows a pattern of following what's iconic in Los Angeles, or, the things that people recognize: Wilshire Boulevard, the Sunset Strip, Hollywood Boulevard, the Kodak Theater. In New York, we see magnet events following similar kinds of iconic infrastructure.

What are event enclaves? They are the places we associate with Los Angeles and New York. In LA, we think about Sunset Boulevard, Hollywood and Beverly Hills. And in New York, we think about Fifth Avenue, Times Square and the West Village. Now, maybe we as New Yorkers don't think of these places when we think of New York, but much of the world does.

EC

Throughout this study, we saw a recursive, reinforcing dynamic in which certain event enclaves—and certain institutions in neighborhoods—tended to have dominance and usurp the possibility of other places becoming centers of entertainment or cultural activity. Some places have a disproportionate importance, and we think this makes sense. It's an economy of scale. Why would the media start showing up somewhere else if they can guarantee getting what they need—the pictures— in a particular place?

The first important finding of our study is that there's nothing random about what we discovered; events do have a pattern, and it's statistically robust. The second is that these different industries aren't hanging out in a vacuum. They're constantly sharing the same spaces and the same scenes, and these places brand themselves. When Times Square becomes Times Square, well, why would you hold an event anywhere besides Times Square?

SW

Getty represents an economic driver, and it helps us to understand economic dynamics spatially. As Elizabeth mentioned, Getty gets out there and looks at what's happening.

EC

Because it's market-driven, Getty depends on the money shot. It's no surprise that the Oscars have so many pictures. Not only is it hugely market-driven, there's a greater chance of catching more celebrities in the flash of a camera, and it's mutually beneficial. The media needs people to photograph at an event in order to have a job. And so, celebrities and cultural industries depend on the media to distribute their products—to get buzz, and to generate buzz in the first place. As glamorous as buzz and celebrity appear to be, these relationships are quite practical. That probably explains why we see such a finite number of event enclaves within a city.

SW

In some ways, settings show that media helps reinforce place branding. We already know that. By taking pictures at Times Square, we further reinforce Times Square as a place to take pictures of, and further reinforce using Times Square as a place to hold events. Media

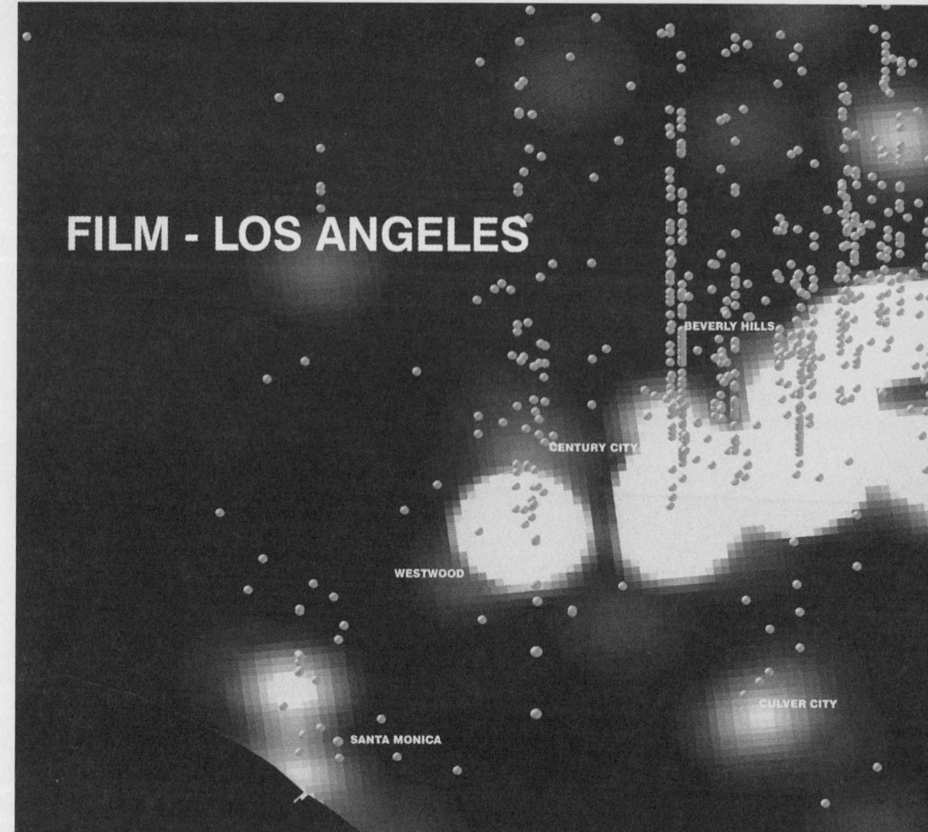

Minna Ninova and Sarah Williams of the Spatial Information Design Lab at Columbia GSAPP created this map of film event enclaves in Los Angeles, based on the Getty Images database.

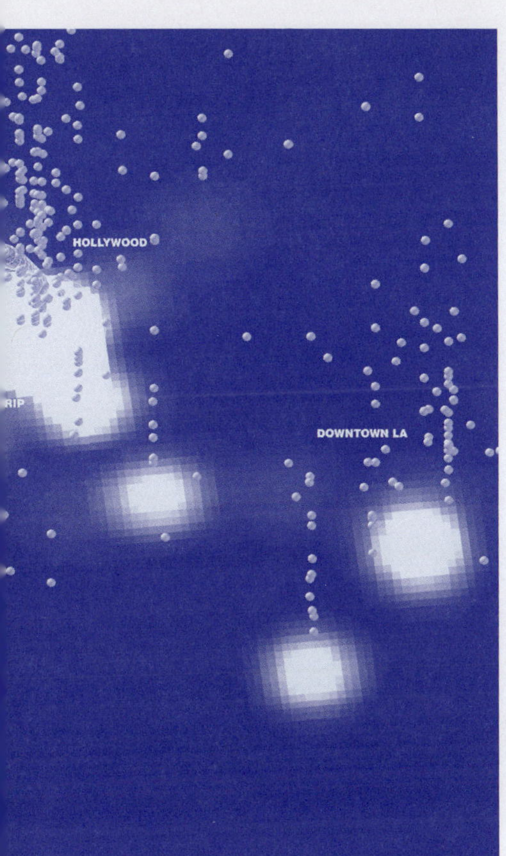

is helping to think about development in the city. If a particular place is hot, it's going to stay hot because the media makes it so.

EC
There's a certain no-duh quality. This isn't telling us something profound. But it is quantifying it in a meaningful way. It does illuminate the tension and the dynamic between the media and creative industries: they depend on each other constantly, and they create winner-take-all markets for particular neighborhoods and event enclaves.

Jamin Brophy-Warren
It's funny, when I first looked at the map I thought, "Where's Brooklyn? Why isn't Brooklyn on here?"

Audience member
Yeah.

JBW
And Elizabeth said it's statistically insignificant. [Laughs]

Audience members
Boo…!

JBW
Anyone who pays rent in Brooklyn knows that that's horribly untrue. [Laughs]
I have a confession. I have a tortured relationship with the word "buzz." This comes from a journalist's perspective. When I started, my editors constantly asked me to find things that were "buzzy," but they didn't know what that word meant. So it served as a proxy for a nebulous notion of coolness. It bothered me because it was a stop-gap. Anytime you read a newspaper article that uses the word "buzz," it's usually because a reporter has pitched an article

to an editor who said, "Why should we care about this?," and they replied, "It's buzzy."

What journalists struggle with, especially those who write about culture, is adding empirical evidence to what we're writing about, as opposed to the disparate stories that we hear. One of the big takeaways of this work for me was being able to quantify this—to be able to tie it to something specific.

It makes me think about how locative technologies affect the way that buzz is tracked. Twitter and status updates are interesting for the ability of a single individual to create buzz around a certain place. Shaq is a huge Twitter guy. He was in Arizona recently, and he showed up at a diner called the Five and Diner. He sent out a tweet saying "I'm at the Five and Diner. Come and meet me." People came and took photos with him. So, the Five and Diner in Arizona... I don't think that Getty was necessarily there. But now it's imbued with a type of buzz that didn't exist before.

I'm also thinking of someone more... "out and about." Lindsay Lohan. Her Twitter feed lists specific places like, "At Chateau meeting." Getty serves as an intermediary between the public and celebrities. But now there is a disintermediation, in which celebrities are able to talk directly to the public.

When I think about buzz, I think about bees. They have one queen, and a few thousand bees. Swarms. In this case, think of Shaq and his hundred and twenty-six thousand Twitter followers. At the end of each summer, bee colonies face this big test. They go and find a new place, and they have to build enough there to house a new colony. That's how I see Twitter. You have the queen bee—Shaq or Lindsay Lohan—and they are able to create buzz in these places.

Erin Aigner

I really like that Sarah's maps are the opposite of traditional cartographic convention, which would have the background be light, and the line work—the data itself—be darker. In other words, they would use ink to show patterns.

I think this is very cool and that it works on multiple levels. For instance, a lot of these events are at night. So to have a map that's totally dark, and then to show data getting brighter and brighter according to the intensity of the event—it's like a flash. It's exactly the metaphor that you want to create for the idea of buzz, and I really appreciate what it evokes. These maps create such great impressions.

I'm wondering if you could hold a constant scale throughout the maps, so you might be able to see how intense the magnitude of different events are? I realize that you then wouldn't be able to compare within the map as often—because if you occasionally have a lower number of events, they're not going to show up—but it would be interesting to see on a more consistent scale.

SW

Definitely. When we were thinking about the maps, we thought "How do we present this buzzy nature, the ideas, the pictures, the photos..." Certainly, we just convey the information—how many events are in the Theater District versus how many are in another location, for example. But ultimately we wanted to ask, "What's a hot spot?"

EA

Right, exactly. I'm in awe of how much data you had to go through and clean.

*I asked my local dry cleaners to keep
any debris brought back by customers.
As the plastic, wire, paper, and
cardboard tubes started accumulating,
it became clear what I needed to do:
create woven structures.*

—Suzanne Tick, 05.14.09

*How bizarre is that?
There are 500 of these birds left in
the world, and there's one
on the ground in Times Square.*

—Kate Orff, 10.15.09

*Zines have always been more than
just words and images on paper.
They're the embodiment of an ethic
of creativity that argues
that anyone can be a creator.*

—Stephen Duncombe, 01.08.09

*When you're viewing a horror film—
your adrenaline—your body is in
a heightened state, and this
[eye-level, pointed, jutting piece
of glass] is my way of entering
that subtle or not-so-subtle state.
You feel like you're here.*

—Heather Rowe, 11.19.09

*There's a lot of ink in architecture,
but it's hidden.
Ink is inside a cartridge, and the
cartridge is put inside a printer,
and the moment that the
ink makes contact with the paper
is absolutely concealed.
This is very similar to a general fear
of liquids—usually a sexual fear.*

—Mark Wigley, 01.28.10

So are we! [Laughs]

EA
Because this method of getting a handle on pop culture is not…. Quantitative data about pop culture is not collected very often. So, just trying to find something that is a proxy, where you can say, "At least for this year, or for this type of event, it's comprehensive…" That's a really difficult task.

It would be great to try it as a time series, to see if there's a shift. You could do it at the beginning of the decade, or at the end. Or you could slice the dataset that you already have, by day of the week, or by season, so you wouldn't have to collect new data. I think you know that this method is great and that there are more opportunities to use what you already have.

Harvey Molotch
This work represents the way that social geography and urban sociology began: by laying out where things are. And out of that came an incredibly rich set of disciplines. What you're seeing tonight is, I think, the beginning of an effort to do that—but with new kinds of phenomena.

These new phenomena stem from an increasing recognition that what moves the world is not tanks and bombs, but human affect: the aesthetics and sensibilities of ordinary life, and the drive to interact. If you treat those kinds of things as making up a coherent whole that lies behind an enormous amount of economic productivity, then you start to see the significance of this work.

This exhibit highlights the zeal for proximity, for face-to-face contact, among people who are in this case celebrities. But I think this moves beyond their world—into the worlds of finance and indeed technology. There, the same phenomena are going on, and meeting face-to-face is critical. The city is where people can dependably run into other people in order to make their projects come into being. It isn't just the media interacting with the public, but members of the public interacting with one another. If you need to go out to a place where you can bump into someone who is appropriate for you to bump into, then there is a short-list of candidate places. There's an economy of sites. Place reliability is important for anyone—it's not just for celebrities working the media. So, yes, in the way you just heard, there are celebrity spaces. But at a lower level of volume, and in a more mundane way, all kinds of places operate as critical— and knowable—meeting grounds for interested parties—people driven by affect as a dimension of life.

This means is that we have to look upon the economy, technology, sensibility and sociability as making up a kind of continuous thread. They're all part of the same dynamic.

Studio-X and Spacebuster

In April 2009, Storefront for Art and Architecture asked Studio-X New York to curate two events within Spacebuster —a mobile, inflatable pavilion by the Berlin-based group Raumlabor.

—

Examined Life with Astra Taylor and Avital Ronell

Astra Taylor, Avital Ronell

Location: Clemente Soto Vélez Cultural and Educational Center, 107 Suffolk Street, New York City

Co-sponsored by and Raumlabor and Storefront for Art and Architecture

April 22, 2009

—

At about 7pm on April 22nd, 2009, nearly 60 people took their seats inside Spacebuster by Raumlabor to watch my documentary, *Examined Life*.

Images of that evening, taken by the intrepid photographer Alan Tansey from the roof of a nearby building, are positively haunting. The Spacebuster bubble, emerging from the back of a beaten-up industrial truck, transforms the vehicle so it looks like some alien life form, a giant translucent urban insect. The sky, captured at that moment when the blue is most dramatic, right as day becomes night, is a tangle of clouds, the light emanating from the tenement windows below so warm it almost seems the insides of the buildings are on fire. If you gaze into the bubble you can make out the glow of the screen as the opening sequence rolls.

The evening was unusually cold given it was already spring, almost frigid. Protected only by the bubble's gossamer skin, the crowd shivered through the film, and many stayed on for a lively discussion with philosopher Avital Ronell, one that became enjoyably heated, however briefly. A movie about philosophy that doesn't spark debate or provoke opinion hasn't done its job.

With *Examined Life,* I wanted to dislocate the space of thought—from universities, libraries, bookshelves—by presenting philosophy in motion through a series of filmed peripatetic excursions. Spacebuster upped this challenge, dislocating the space of cinema itself, taking the movie out of the theater and into a parking lot on the Lower East Side. Inside the inflated dome the sounds of the film—so painstakingly mixed in a studio, augmented by a week of Foley effects sessions—were lent a new dimension through the din of the real street and the reverberations of traffic nearby. As a reward for submitting to the elements—the unaccommodating temperature, the wait for the sun to descend so the projection was visible, the noise of New York City—we had a sense of being in the world amidst all its chaotic, uncontainable glory.

—Astra Taylor

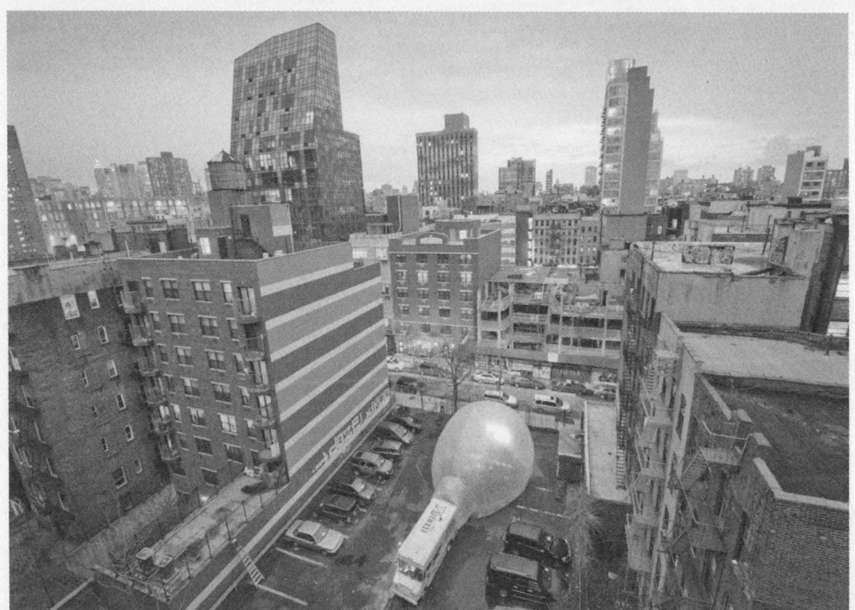

Lower East Side

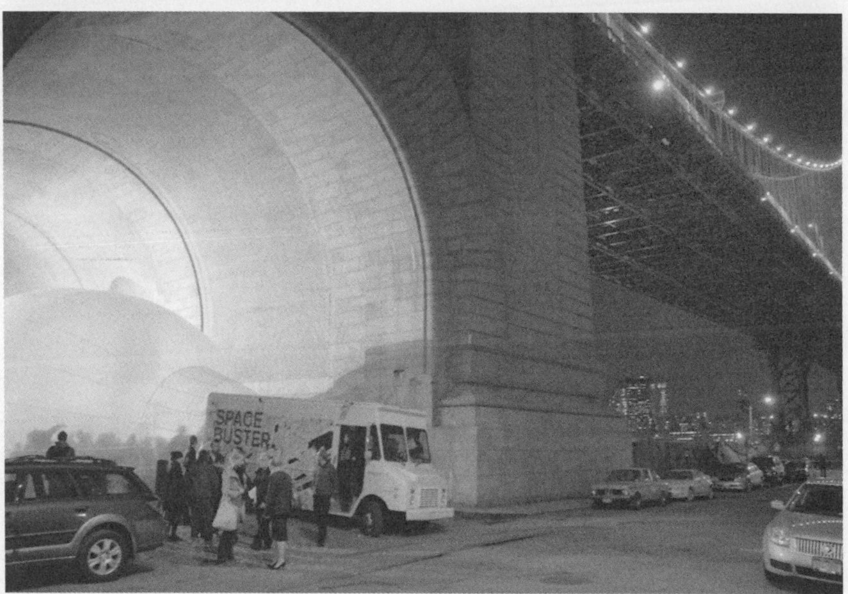

DUMBO, Brooklyn

Co-curated by Mitchell Joachim and Ioanna Theocharopoulou, ECOGRAM is an ongoing conference that addresses ecology, sustainable design, planning and development. This diagram outlines the various ways in which ECOGRAM and Studio-X New York have collaborated and informed each other thus far, from the first conference in 2008 through IRON DESIGNER (April 2009) and the *Safari 7 Reading Room* (October–December 2009).

—

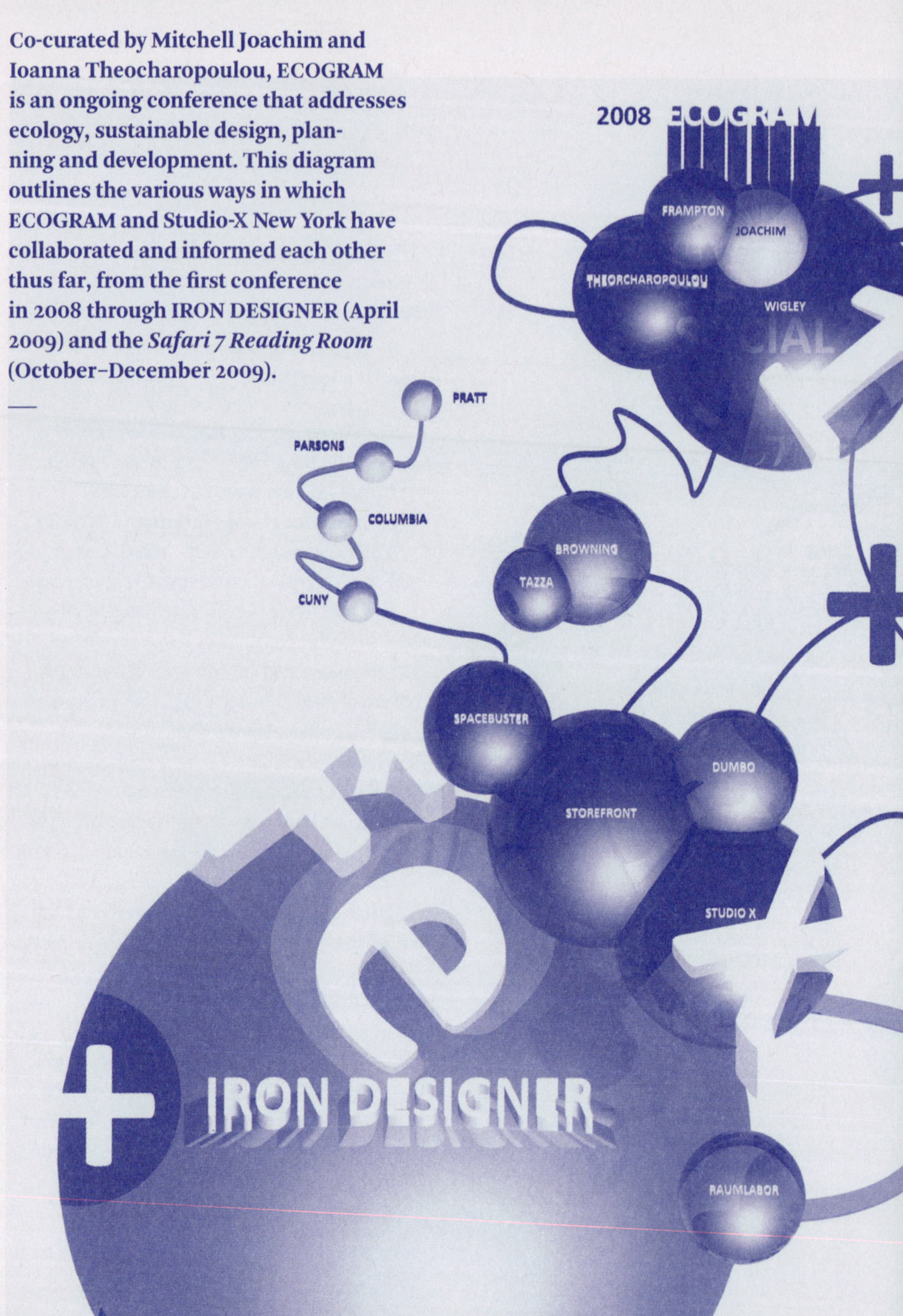

IRON DESIGNER

Gavin Browning, Mitchell Joachim,
Ioanna Theocharopoulou

Location: The Archway, DUMBO,
Brooklyn

Co-sponsored by the DUMBO
Improvement District, *Inhabitat*,
Raumlabor and Storefront for
Art and Architecture.

April 23, 2009

Find *Tazza* in the diagram on the previous page. It's a cafe in Brooklyn Heights where Mitchell Joachim, Ioanna Theocharopoulou and Gavin Browning met and hatched the idea of an ecological, site-specific, participatory, live design competition: IRON DESIGNER. Based on *Iron Chef*, the event was held inside Spacebuster by Raumlabor. This bubble was anchored for one night inside the Manhattan Bridge's Archway—a cavernous space closed to the public for two decades and opened by the NYC Department of Transportation for this occasion.

Inside these spaces, teams of Third-Year M.Arch students from City College of New York, Columbia GSAPP, Parsons The New School for Design and Pratt Institute were handed a sealed envelope containing a site-specific design challenge/recipe, and were given one hour to complete the task. IRON DESIGNER was a party, a spectacle and a happening rolled into one—signs of life and creativity not seen in this corridor of Brooklyn for twenty years.
—

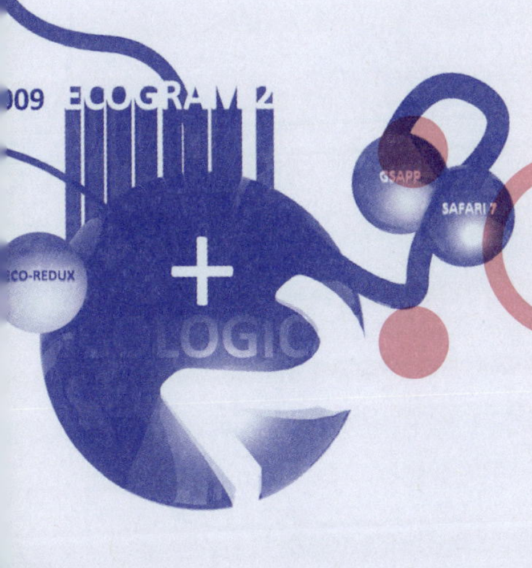

IRON DESIGNER Recipe

Active ingredients:
• The Lot
• The Archway
• The Missing Link(s)

Instructions:
This challenge calls for the enhancement of public space in DUMBO, Brooklyn, and for a contribution to the ecological city. The Archway under the Manhattan Bridge has been closed to the public for the last seventeen years. It has just reopened, and plans call for it to accommodate various public programs including a market, a community theater, and boxing matches. The Lot directly adjacent to the Archway—currently fenced off and being used as a storage space for transportation and building materials— needs an identity in order to make it a viable public space.

   Your job is to create a link between the site's two anchors, the Archway and the adjacent Lot, by suggesting a new program of your own invention, and by developing it as much as possible within one hour. The limits of the site are variable. It can extend in concept towards Fulton Ferry, the Navy Yard or across the East River to Manhattan's Chinatown.

Some possible PUBLIC programmatic choices/mixes include:
• A market
• A vertical farm
• A skatepark or extreme sporting arena
• Noise-reduction systems and pedestrian safety
• A new multi-nodal transportation link or gateway to the river, or to other parts of Brooklyn

Please consider multiple users: age, ability, income, ethnicity.

You may use whatever format you choose to attack this challenge.

Judging Criteria:
1. Team attitude and school spirit
2. Sustainability, broadly defined: social, economic, environmental
3. Context sensitivity: how well it serves the public(s) of New York City
4. Novelty
5. Craft and presentation

Columbia University, GSAPP
Troy Therrien, Kyung Jae Kim, Christopher Barley, Marlo Brown

"Columbia's playful design proposed creating sculptural, deployable balls of various sizes to sit in the Archway. Each ball contains the necessary materials for a different activity, anything from flotation devices to picnic benches. Intended to encourage civic engagement, the balls need the public to activate the programming."

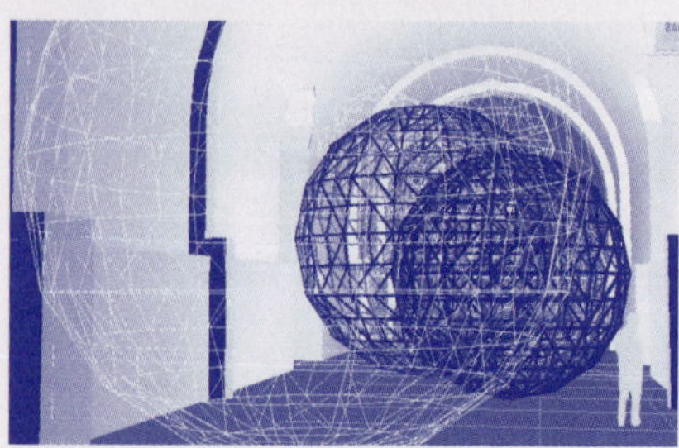

Pratt Institute
Adrien Allred, Renee Glick, Hart Marlow, Elliot White

"Perhaps inspired by Spacebuster's unusual acoustics, the Pratt team built their design around the idea of creating vibrations. The design symbolically collects 'sediments' of human interactions in the Archway—with a particular focus on the vibrational energy created by occupiers of the space. The team proposed a multi-level structure that would provide numerous flexible spaces appearing to extend out of the Archway—just as vibrations would also extend out of the space."

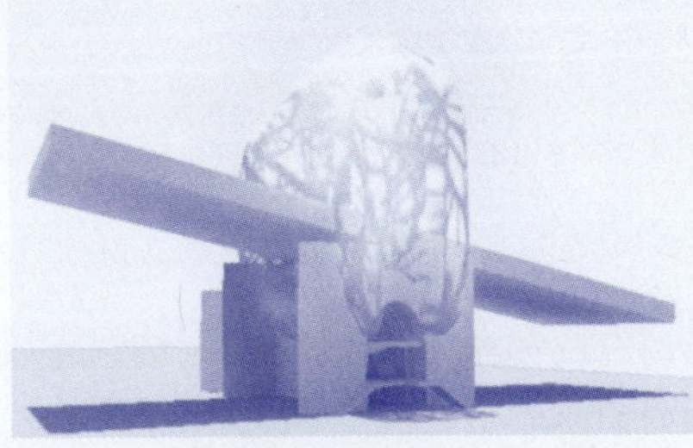

Parsons The New School For Design
Hrolfur Karl Cela, Margot Otten, Kristen Teutonico, Matthew Bissen

"The Parsons team created a two-pronged, connected program: a market and nighttime dining table. The team proposed using the adjacent lot as a farmer's market and then creating a dining table for the community to gather under the Archway. Food waste created by diners would later be composted and used to grow more food."

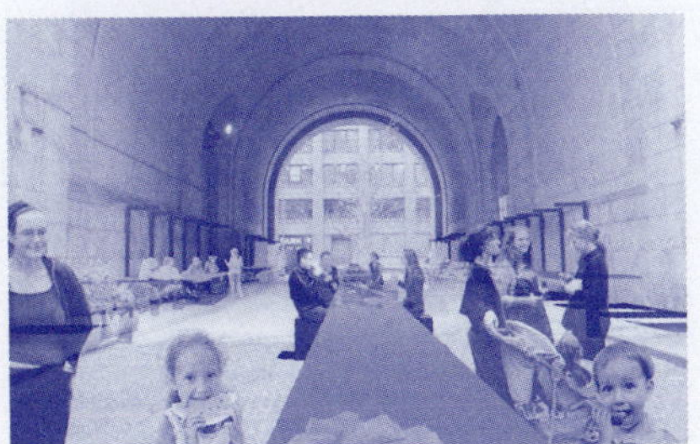

City College of New York
Halina Steiner, Brett Seamans, Perry Randazzo, Orlando Rymer

"CCNY proposed a flexible space that could be used by neighboring artists and galleries. They designed a fluid canopy, suspended by cables, to cover the adjacent lot. Seating in the Archway is designed to mimic the shape of the canopy, in order to create a uniform appearance. The Archway would be flanked by bioswales to collect stormwater run-off from the bridge, creating landscaped spaces for a more pleasant environment."

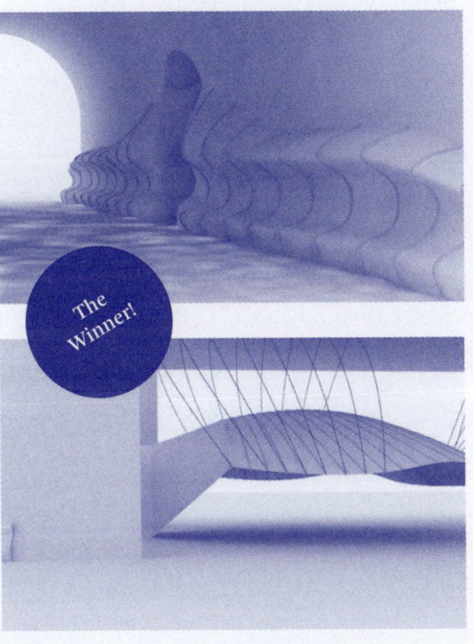

All quotes from Olivia Chen, "IRON DESIGNER: The Winners!" *Inhabitat*, 4/25/09

Glen Cummings, Janette Kim, Kate Orff

October 15–December 31, 2009

Led by the Urban Landscape Lab (Janette Kim, Kate Orff) and MTWTF (Glen Cummings), Safari 7 began in the summer of 2009 as a series of podcasts and maps that explored the ecosystems and wildlife habitats along the route of the MTA No. 7 train. On the occasion of ECOGRAM 2, this self-guided tour was then transformed into an exhibition at Studio-X New York: a "reading room" featuring 3D maps, wall graphics and a curated selection of books, where New Yorkers could learn about the rich ecology all around them. Oysters, dogs, humans, worms, snakefish, cormorants and germs are just some of the species that populate this urban transect. Recently, *Safari 7 Reading Room* moved further into the public realm. It was displayed in Grand Central Station's Vanderbilt Hall for Earth Week, 2010.

—

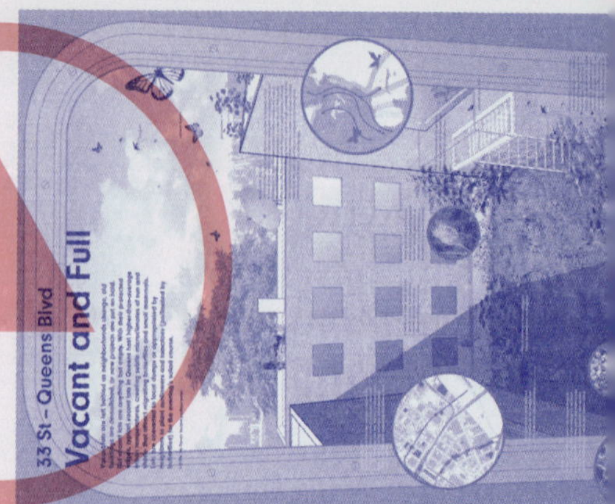

Vacant and Full

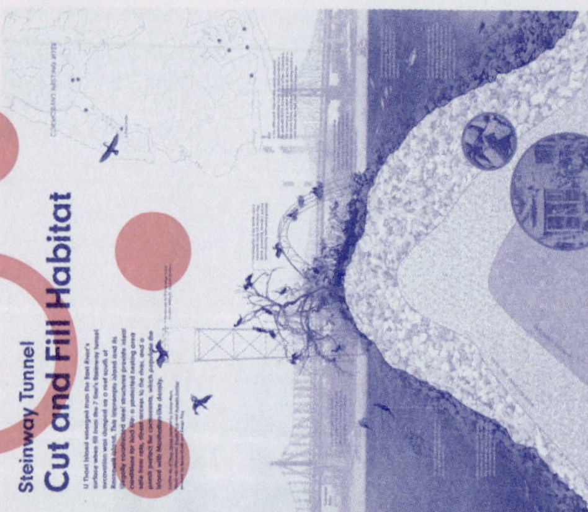

Cut and Fill Habitat

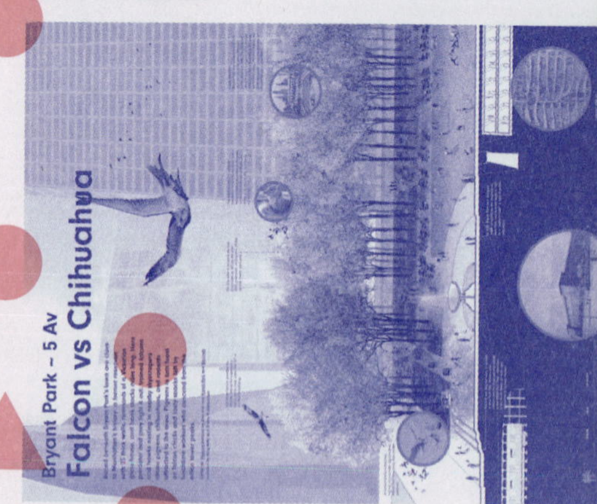

Falcon vs Chihuahua

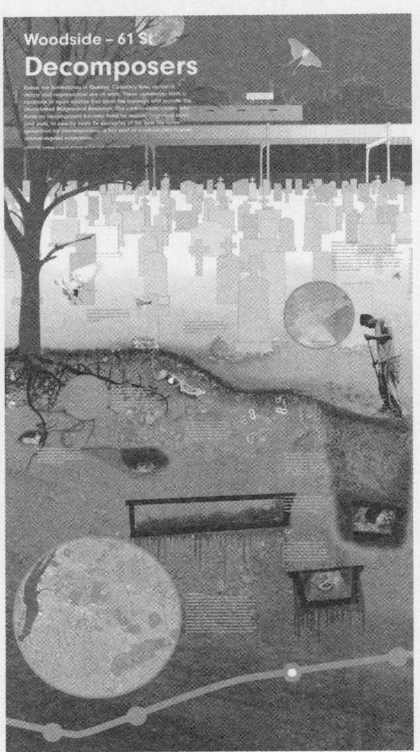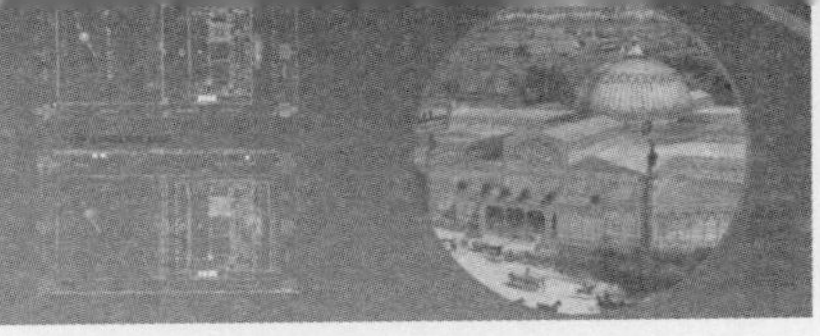

Decomposers

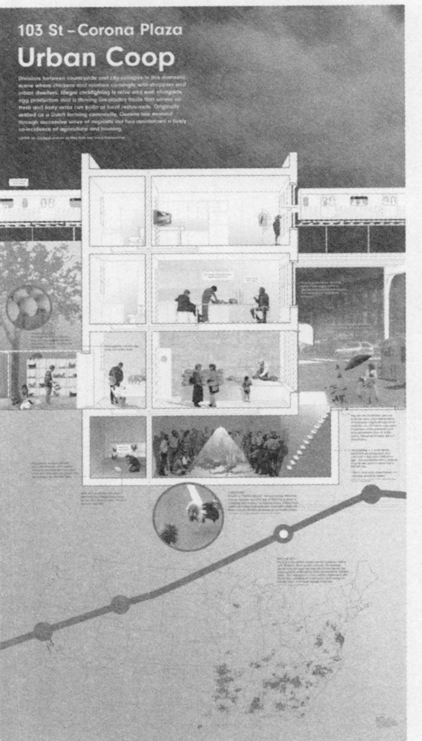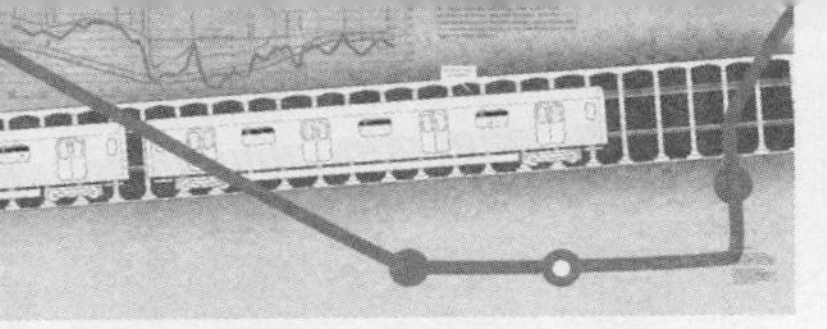

Urban Coop

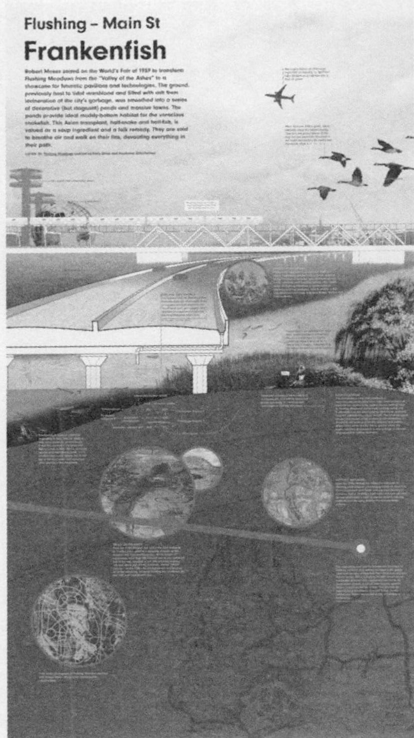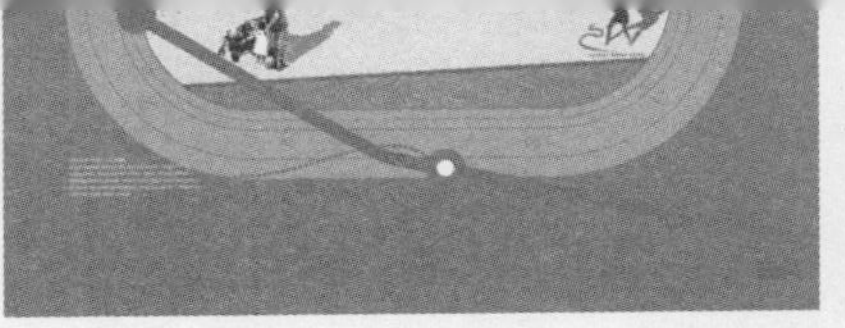

Frankenfish

<u>**Imagining Recovery**</u>

Wayne Congar, Troy Therrien

May 13, 2009

After designers Wayne Congar and Troy Therrien attended the January 2009 event *Rapid Response: Addressing the Address*, they devised this international design competition to envision future economic recoveries via the Recovery Act, as quantified by the charts, graphs, maps and accounting figures detailed on the US government website *recovery.org*.

—

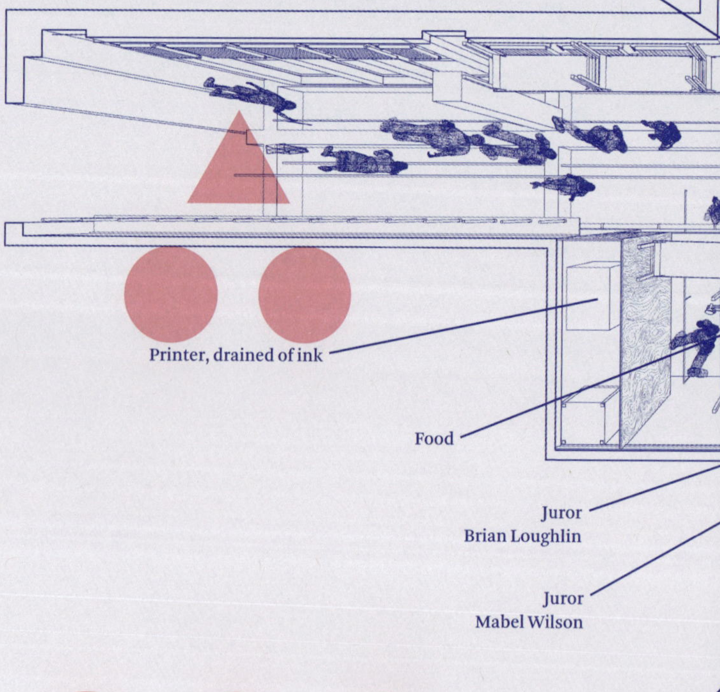

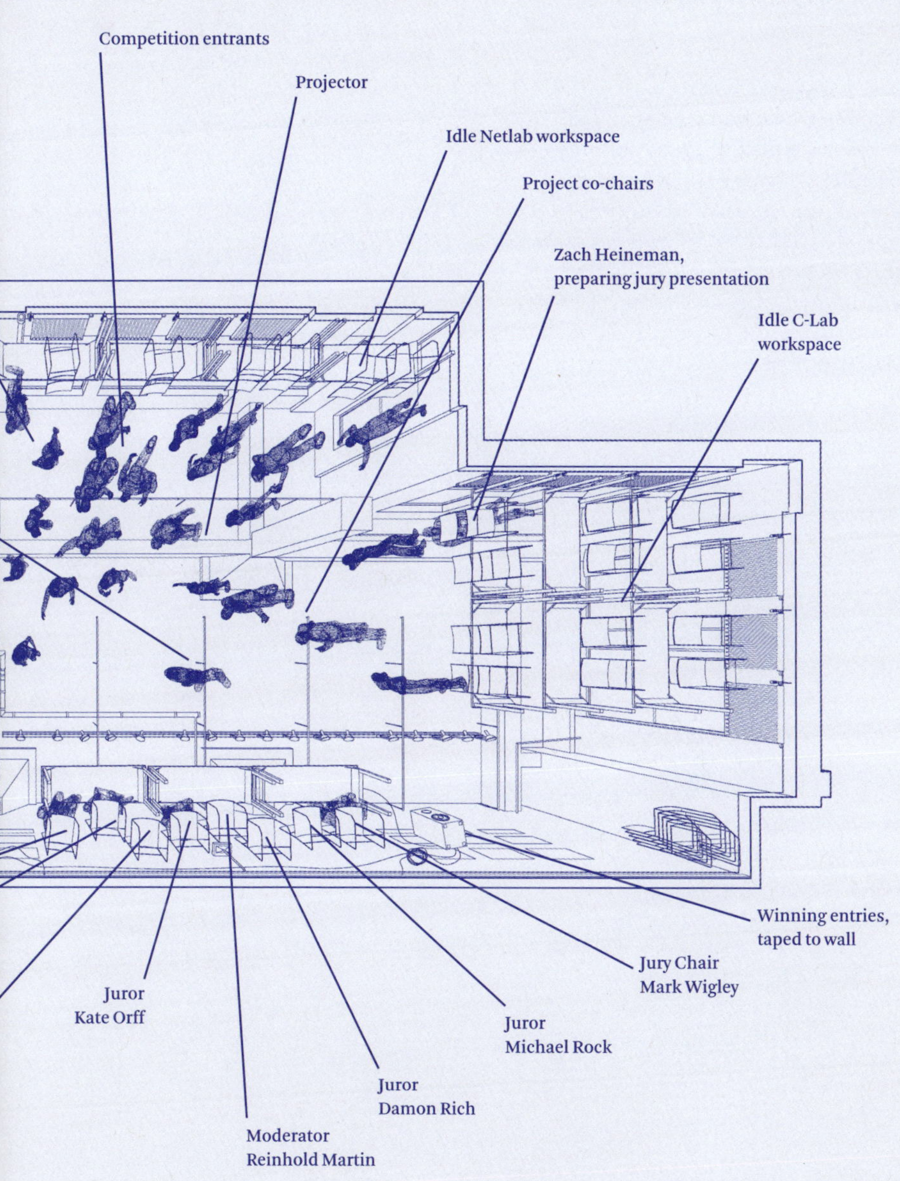

Competition entrants
Projector
Idle Netlab workspace
Project co-chairs
Zach Heineman, preparing jury presentation
Idle C-Lab workspace
Winning entries, taped to wall
Jury Chair Mark Wigley
Juror Michael Rock
Juror Damon Rich
Moderator Reinhold Martin
Juror Kate Orff

InDisposed: Talking Trash about Design

Curated by Jen Renzi and Dan Rubinstein

Featuring: Ate Atema, Tom Chiu, David First, Kegan Fisher, Liz Kinnmark, Adrian Kondratowicz, Paul Loebach, Kevin Patrick McCarthy, Jeff Miller, Takeshi Miyakawa, Andrea Ruggiero, MIO, Situ Studio, So Takahashi, Suzanne Tick, Alex Valich, Christine Warren, Tobias Wong

May 14–21, 2009

An offsite exhibition held during the 2009 International Contemporary Furniture Fair, *InDisposed: Talking Trash about Design,* according to its curators, "highlighted the twin dichotomies that define contemporary design today: sustainability versus wastefulness, and preciousness versus mass production." Renzi and Rubinstein challenged designers to create pieces from environmentally friendly materials that could be used, possibly mass produced, and then disposed of in a responsible way. The results? A curtain woven from leftover dry cleaning materials, a table that eats itself, a pile of pink, peppermint-scented garbage bags that challenge the invisibility of waste in our urban landscape, and more. After debuting at Studio-X New York, the show traveled on to Los Angeles' Touch Gallery.

—

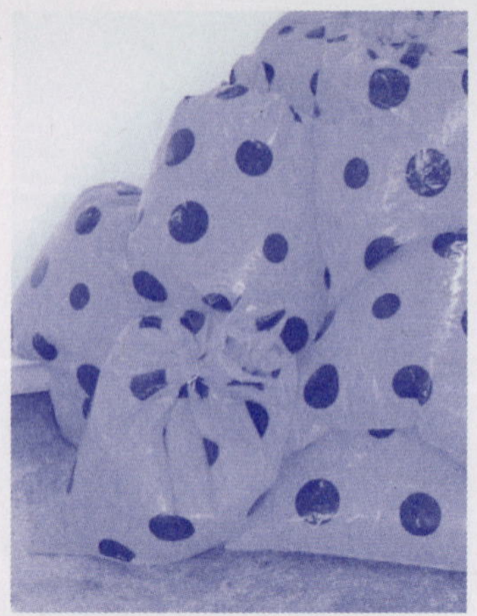

Adrian Kondratowicz

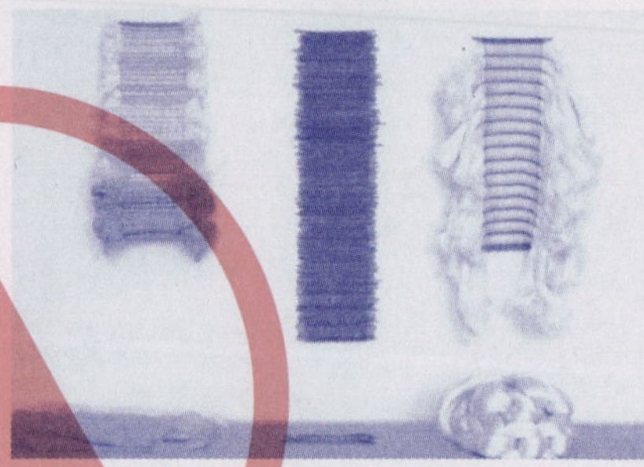

Suzanne Tick

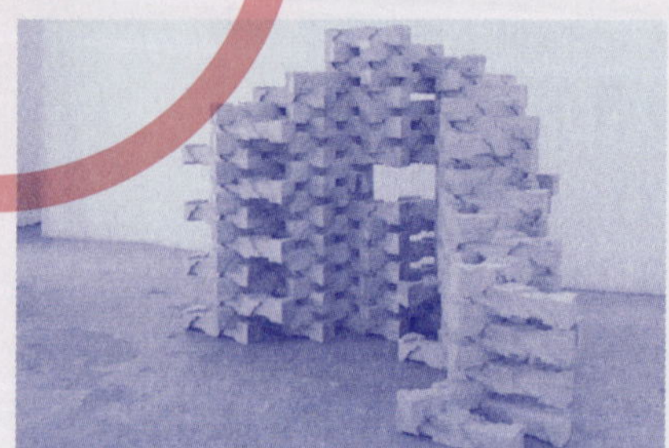

Situ Studio

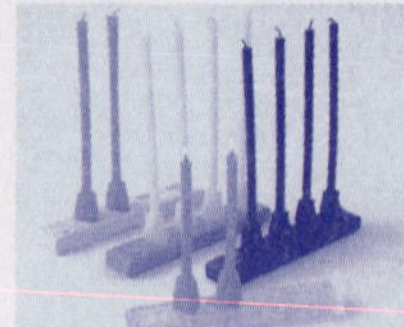

Design Glut

I did a calculation a while ago about executive pay of senior executives over the last forty years, by which I mean the five main officers whose pay was disclosed. If the stock market grew at the same rate as their pay, the Dow, which has been hovering for a long time around 10,000, would be at 80,000.

—David Cay Johnston, 02.11.10

At one time the tools needed for making a zine were glue and an X-Acto knife; now, their digital counterparts are second nature to architects.

—Mimi Zeiger, 01.08.09

*Often when I walk in the
Ramallah hills, I try to avoid seeing
the destruction caused to
the landscape by Israeli settlements,
roads, and the Wall.
Instead I concentrate on looking
at what remains unscathed. This time,
I could not be so discriminating.*

—Raja Shehadeh, 03.22.10

*Monuments and memorials literally
re-member. They bring together;
they re-collect. This ritualistic
reenactment of drawing together again
becomes collective. It becomes
the process of
remembering and recollecting.*

—Mabel Wilson, 11.10.09

Reinhold Martin, Diana Martinez, Leah Meisterlin

June 12, 2009

The Temple Hoyne Buell Center for the Study of American Architecture convened a day-long policy and design workshop on public housing at Studio-X New York. Attended by over forty faculty, students and recent graduates of Columbia GSAPP, the workshop and its results became an Autumn 2009 exhibition at Columbia's Avery Hall, a public conversation, a tangentially related symposium on "architecture in public," and a punchy, informative pamphlet, *Public Housing: A New Conversation*, discussed here by Buell Center Director Reinhold Martin and *GRITtv*'s Laura Flanders on December 17, 2009.

—

Laura Flanders
For some people, it might come as a surprise that the US still has public housing.

Reinhold Martin
Yes, we seem to have forgotten the first fact about public housing in this country: it exists. This is a self-evident premise, but that it exists is part of the reason that we don't talk about it today.

LF
Does it exist in one form? I think the public's impression is that those buildings used to be public housing projects, but now they're something else.

RM
It exists most concretely in the public imagination. After decades of turning this question around in the mass media, in literature, in all kinds of ways, "the projects" are allowed to stand for what is in reality a far more complex, much more textured and nuanced aspect of the American landscape. And not just in cities—that's an important aspect.

LF
Is that to say that housing projects stand for something sad or negative in American popular culture?

RM
Yes, they're stigmatized, and the stigmas are not entirely related to the actual housing complexes. Public housing— a very real figment of the American imagination, connected with very real objects, people and lives—condenses a whole set of social issues around the city. Racial, gender and economic disparities of all kinds.

LF
Was it always this way? If you go back to the early part of the last century, people were excited that public housing was being built. It was an improvement over what existed before.

RM
Much of the history of public housing in the US tracks back to the New Deal and some of the policies and projects that emanated out of that.

LF
What was the thought about public housing at that point?

RM
The *thought* itself was actually

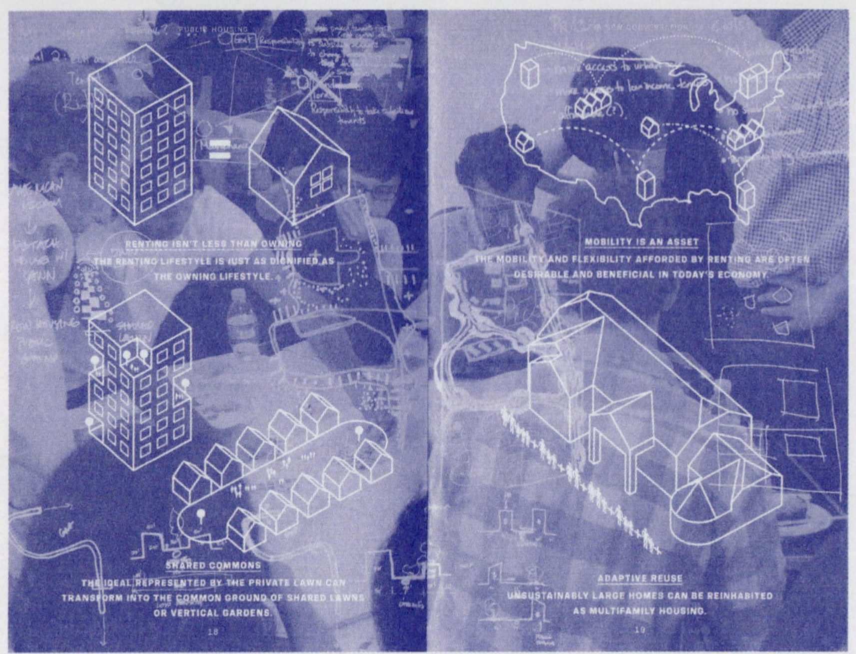

From Charrette to Publication

complicated, because what emerged was the idea that what we now call "public housing" was a means towards another kind of an end. That is, home ownership. The story of public housing in this country is a story of forks in the road. A series of decisions were made that were organized around political questions. On one hand, these were questions of affordability. But they were also organized around what we can simply call "dreams." The so-called American Dream has organized the narrative of public housing from the beginning, in that many of these complexes seem to be temporary places where people would live for a while until they could afford to buy a home in the suburbs. That narrative—that dream—is part of what we need to work on, because dreams can frame actual policy, and they have framed decisions at various historic forks in the road.

LF

If you want to change the dream, or the trajectory that we associate with that dream, then how would public housing be different?

RM

It would begin by recognizing that all housing is public. Federal and state governments subsidize all forms of housing—from home ownership to what we generally consider public housing—through tax credits or direct subsidy. The idea that there is a thing called "the public," and that "the public is not necessarily me" is part of the stigma, and it is the first hurdle. The public is what is common, but new social movements and other kinds of political processes have also taught us that the public is *multiple*. There are many publics and they are not reducible to their self-interests. Importantly, whatever one would describe as the "public sphere"—the space in which shared concerns are debated, discussed and formulated, and the space in which dreams are produced—is where this set of publics is comparing notes. It's constantly in the process of debating its priorities.

LF
After Hurricane Katrina and the breaking of the levees, local people in New Orleans were told: "We're just not going to rebuild public housing." Even public housing that *wasn't* damaged in the storm was demolished. How do you go back into that argument?

RM
You have to get behind it. You have to ask, "What are the premises under which such a statement is possible?" The public sphere—in whatever shape or form it takes—makes certain things conceivable and certain policy avenues available. Likewise, it takes others off the table. Public housing has been so thoroughly framed in stigmas that that statement seemed self-evident to many people. Frequently, these statements conceal and reveal vested interests. In this case, the statement was made in relationship to the interests of the markets.

LF
You make the point that housing is a verb. It has to do with living and what happens in homes.

RM
Yes, housing is a process of negotiation, discussion, debate and disagreement. For example, many public housing complexes were built in a top-down way: state representatives, architects and planners more or less superimposed their models on cities. In many cases, these models have not worked. But there are certainly many new, interactive, conversational, even agnostic models that are available, and debates can be had based on disagreements and conflicts of interest. That's what publics are. Publics are necessarily in disagreement.

LF
How do you reconfigure our idea of the public? Particularly, you talk about individuals. You say, "Individuals and groups live along a spectrum of collectivity." Why does that matter?

RM
It matters in the sense that you are able to imagine yourself as a constituent in a public sphere.

LF
And that's positive? [Laughs]

RM
Yes, that's positive. [Laughs]

LF
Because the culture presents a trajectory of getting out of the project. Escaping the public, and going into the private— the beautiful, private realm that has moral connotations of upstanding citizenship, the suburbs and all the rest.

RM
Yes.

LF
You're saying, let's go back, and let's look at the "public" part. The part that government is always a part of, and that all people are a part of. And let's change attitudes toward *that* part. That's huge.

RM
Yes, it is huge. But it's very, very important. There are plenty of examples in
which this stigma does not hold. Public
universities, for example, or public
facilities, like hospitals. Each has its
own challenges, fiscal or otherwise. But
the values that frame the relationship
between government and its citizens—
or more broadly, between governments
and markets in relation to citizens—
those sets of values need to be elaborat-
ed. They need to be frankly put on
the table, and they need to be debated.

No-Body Zone: Designing an Exploded Audiobook

Part I: Greta Byrum, Stephen Mosblech, Danny Snelson. Moderated by Jason Zuzga.

Part II: Live performance of *Eastern/Western Meat Lessons* featuring Arthur Burkle, Sarah Dahlen, Jay Smith, and Danny Snelson. Written and directed by Stephen Mosblech.

July 14, 2009

The No-Body Project emerged as a med-itation on the evolution of the book—three audiobook producers adapted a non-linear text for new media produc-tion. The result was a kind of hybrid archive/audiobook/collage of Richard Foreman's 1997 novel-in-parts *No-body*. The project features a graphic user interface which allows users to combine elements of audio, video, and text into their own reading experience. When producers Greta Byrum, Stephen Mosblech, and Danny Snelson launched the project at Studio-X New York, a debate ensued about the viability of the current publishing industry and ways in which it could adapt to meet the demands of a new kind of readership, raised on non-linear, user-generated information sharing.

A play by Stephen Mosblech, de-signed for the Studio-X New York space, followed. It was recorded and added to the No-Body archive, along with the script, which is reproduced here.

Eastern/Western Meat Lessons
by Stephen Mosblech

LEO/Jay Smith
MARIE/Sarah Dahlen
SAMUEL/Danny Snelson
ATTENDANT/Arthur Burkle
DEEP VOICE [voice of]/Richard
    Foreman
FEMALE VOICE [voice of]/Juliana
    Francis-Kelly

Sound Design: Greta Byrum
Videographer: Eduardo Band

NOTE: underlined text excerpted from
No-Body: A Novel in Parts by Richard
Foreman

LOCATION: A country road a stroll
from the city. Pasture of AstroTurf
center stage and blazing flat rectangular
orange sunset. Bardo tube off in right
corner collecting dust.

—Scene 1—

   Deep Voice
This has a lot to do with Samuel
This has a lot to do with Samuel
This has a lot to do with Samuel ·

   Leo
Fuck me. Shit for brains.

   Marie
No. I just pissed myself. No.

   Leo
OR Tell me shit-head
who isn't dead now who didn't die?

   Samuel
Experience is limitless that is
Samuel entered the milky abyss

For an instant perhaps:

   Marie
No. I have indigestion. I have
never been so sick in my life Samuel
with worry.

   Attendant
What good does that do.

   Leo
I know I have loved too much
stuffed too many bodies.
[pause]
Hardly a day passes.

   Marie
No.

   Leo
What.
What goes in comes out:
And no pictures

   Marie
Wait wait
since my head is flat I forgot to say
get that rock off stage since it is
shining but it had existed
Pancakes.

   Leo
We all dreamed it together. Wheee!

   Attendant
We all dreamed it together.

   Marie
We all dreamed it together.

   Female Voice
Thank you. Thank You.

   Drone

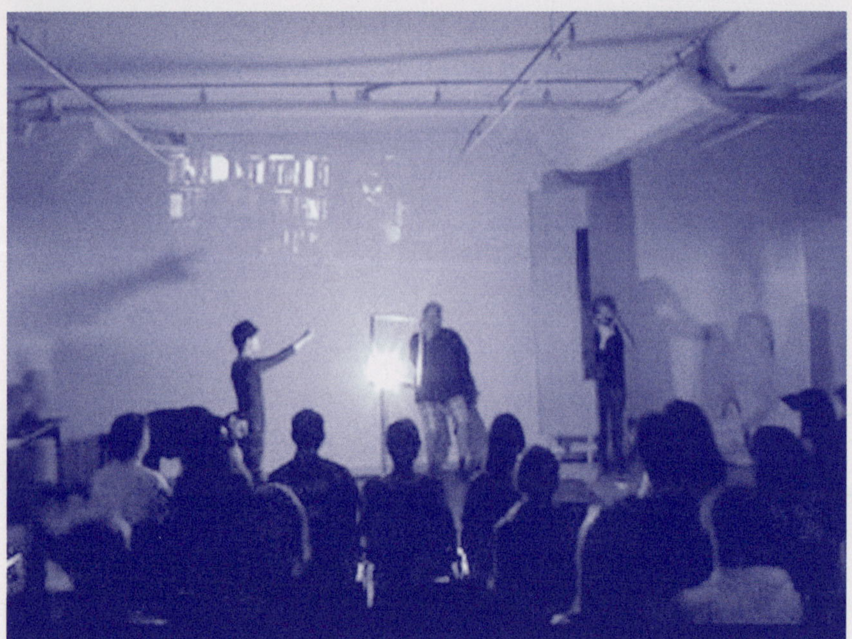

Premiere Performance

—Scene 2—

   Deep Voice
This has a lot to do with Samuel.
This has a lot to do with Samuel.

   Marie
Ach Marie its true—the world is
indigestible In my other hand at
this instant in my hand
elevating a glass of santorum.

   Marie
One, I don't have legs, Samuel
Two, nothing moves plus I can't pin-
point my legs.

   Leo
Does it—this so called no outer and
inner moving—Marie
frighten you.

   Marie
No.

   Leo
Oh Marie
from in this body, the size of the uni-
verse you couldn't at this instant
gonna be picturing the two mummum-
Father degenerations upon the wall
that abducted Samuel at birth unless
Samuel shat god-forbid my brains
on the floor at lunch yesterday and it
wasn't a rock.

   Samuel
I didn't do it.

   Leo
Is your brain perhaps the color orange.

   Samuel
I didn't do it.

Marie
Oh. Oh.

Leo
Discover, Marie
Where a releasement
When I elevate my hand
Spurted from the back of my skull.

Marie
Careful Samuel your teeth'll fall out.

Female Voice
Thank You. Thank you

Samuel
<u>Two things were available to Samuel.
The city, through which he walked.
The language—that was not walked
through but did manipulate his steps,
his direction, his speed.</u>

Drone

—Scene 3—

Deep Voice
This has a lot to do with Samuel
This has a lot to do with Samuel

Attendant
A second object.

Samuel
<u>Is it my intense demeanor that should
convince you to take each of my ideas
seriously. No you see there's nothing
intense about me.</u>

Leo
Is it my intense demeanor that should
convince you to take each of my
ideas seriously. No you see there's nothing intense about me.

Marie
What. What. Wait.

Leo
The Faces.

Marie
No. Look.
Samuel entered the milky abyss

Attendant
Again again again again

Marie
Samuel is chasing me through the
streets with a knife.
I keep trying to stand but I. keep.
Sitting down.

Leo
Oh look. nothing to say about that
i.e. expanse of wings, Cloudlessness
suddenly my body feels heavy
maybe I'll lift myself up from sloth
and d-pression the only way how.

Attendant
I didn't do it.
I didn't do it—
(sings) tomorrow.

Samuel
<u>You Started this Marie</u>

Marie
<u>Oh no Samuel, you started this</u>

Samuel
<u>How</u>

Marie
<u>You wanted it to happen Samuel
Your life as it will be</u>

Attendant
Like lightning Like lightning

Leo
But at that instant a fire alarm bell was
ringing in the distance
and someone else, not Samuel, said:

Marie
Chinee Japanee let's go to Paris
On my magic carpet to get out of here.

Leo
When I look at the city with my hands
from above
It just keeps getting reflected

Female Voice
Thank you.
Thank you.

Samuel
You know what I'm doing? I'm
standing in front of an open window
and the curtains blow and I'm not
moving just like that.

Marie
Eureka
Eureka

Dispatches from Villa Feuerlöscher

**Barbara Hirnthaler, Gabriel Hirnthaler,
Karen Finley, Hannes Priesch.
Response by Mabel Wilson.**

November 10, 2009

**Bringing Austria to Varick Street, Karen
Finley discussed the influences and
intentions behind *Open Heart*. Her
piece is a Holocaust memorial created
through participatory process in New
York City and installed at the historic
Styrian residence Villa Feuerlöscher as
part of a site-specific summer 2009
exhibition about memory, activism, and
artistic practice. The project commemo-
rates 420 children who were murdered
by lethal injection to the heart on a
single day in February 1945 at Gusen
Concentration Camp. Some of the small
clay hearts that comprise *Open Heart*
were made in workshops that Finley led
with elderly Holocaust survivors and
with children, and are now being trans-
ferred from the villa to be permanently
installed on the Gusen grounds.**
—

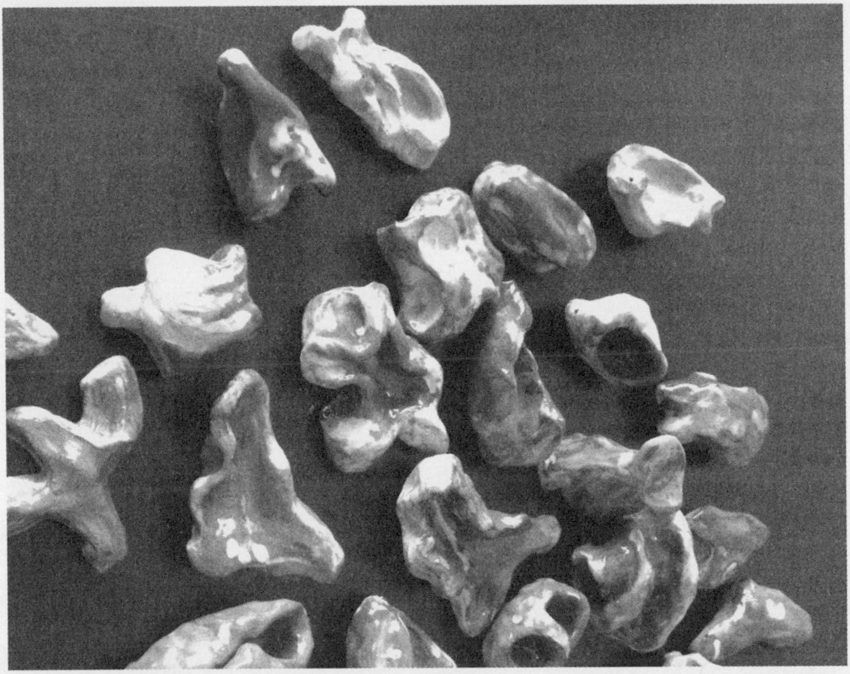

Clay Hearts

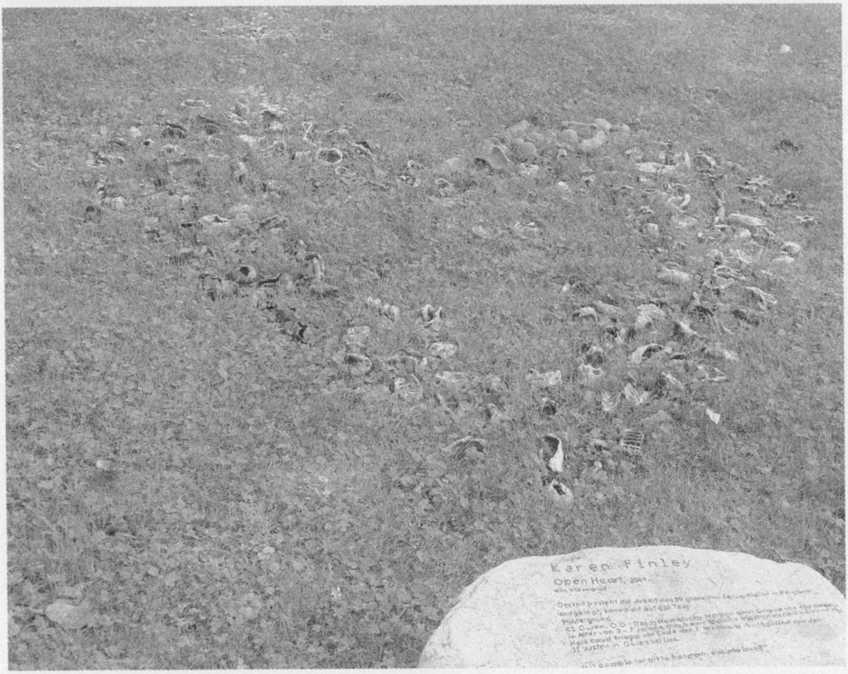

Open Heart at Villa Feuerlöscher

Red Lines Housing Crisis Learning Center Report at Studio-X

Larissa Harris, Prerana Reddy, Damon Rich. Co-sponsored by the Queens Museum of Art

September 17, 2009

This event reported back from *Red Lines Housing Crisis Learning Center*, which was shown at the Queens Museum of Art in the summer of 2009: an exhibition of graphics, models, videos and archival materials that explores home finance from the Great Depression to the Subprime Meltdown. It was presented by those key to its success: artist and urban designer Damon Rich; curator Larissa Harris, who originally commissioned the show at the Center for Advanced Visual Studies at MIT, then expanded it to map 2008 foreclosure filings onto the QMA's famed *Panorama of the City of New York*; and QMA Director of Public Events Prerana Reddy, who organized town hall meetings in high-foreclosure neighborhoods in Queens as part of a broad array of arts-oriented programming that re-imagines the museum as a space for advocacy, community development and social change.

—

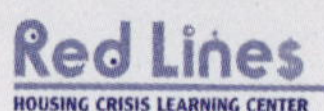

Red Lines Town Hall Meeting

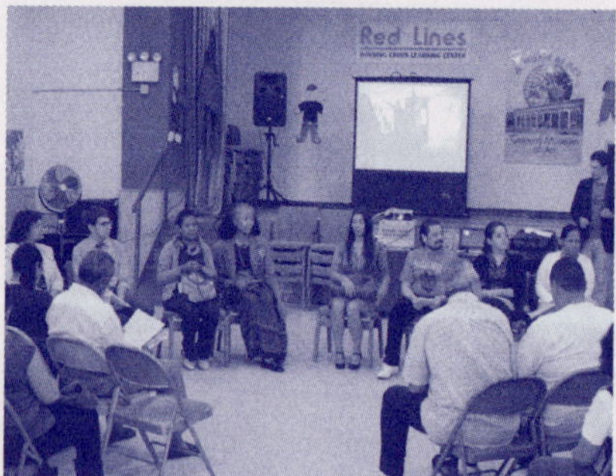
Queens Museum of Art in the Community

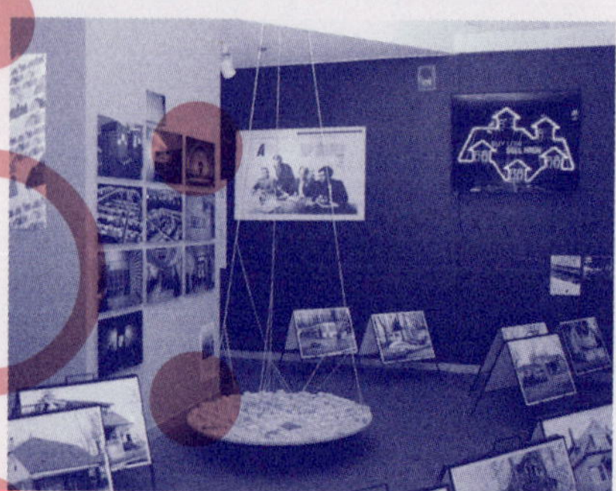
Red Lines at MIT

*My desire to see the incredible
ramps of the Centrosoyuz
led to my arrest in the Soviet Union.*

—Barry Bergdoll, 12.02.08

*The White Nightshade can make its
own seed—without bees—and that's
probably how they're making
it here in these sidewalk cracks,
next to this really nice BMW.*

—Steven Handel, 04.22.10

*Architecture is a public thing, right?
If you want to engage
the world, build a building.*

—Ashley Schafer, 12.04.09

The Garment District

Magda Aboulfadl, Barry Dinerstein, Stan Herman, Patrick Murphy, Barbara Randall. Moderated by Vishaan Chakrabarti.

November 17, 2009

What is the future of New York City's Garment District? Representatives of different and oftentimes at-odds stakeholder groups came together in a public forum to discuss a proposed rezoning of the district. They addressed the changing market pressures faced by New York City's fashion industry, and the infrastructure that keeps it vital. Due to the sensitivity of the topic, and because this was only the second time that representatives from each interest group had gathered in the same space, a neutral environment and an informed moderator were essential.

—

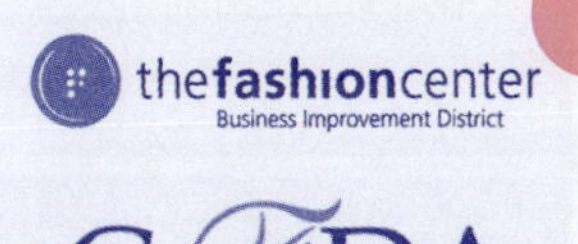

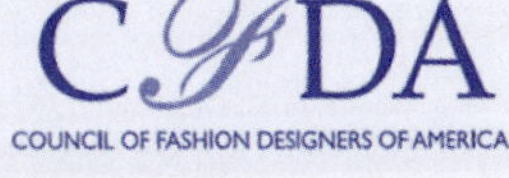

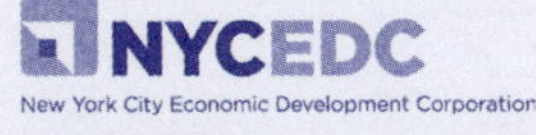

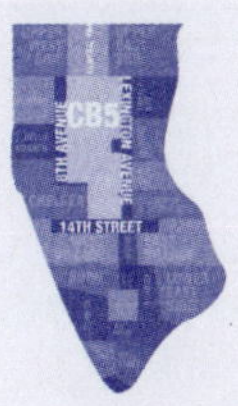

ink

Michelle Fornabai
In New York: Michelle Fornabai, Jonas Mekas, Mark Wigley
In Beijing: Qing Pan, Eric Xu

January 28–March 23, 2010

***ink* is more than just a collection of ink paintings, ink on mylar drawings, and mock material studies in concrete by Michelle Fornabai. It is a series of conversations between Chinese and American designers over the course of one year. It also gave rise to the first real-time simulcast conversation between Studio-X New York and Studio-X Beijing, in which the artist was joined by an international cast of thinkers. Simultaneously evoking American Abstract Expressionism, Rorschach testing, Chinese calligraphy and a newly minted architectural genre, *Concrete Poetry,* Fornabai's haunting and beautiful ink research moved Studio-X New York into new geographic and psychic territories well beyond its Varick Street home.**

—

Michelle Fornabai
The *ink* dialogues are intended to be more provocative than exhaustive, occurring over the course of one year between Beijing and New York City. *ink* exhibitions are held in conjunction with *ink* events.

The first *ink* panel in Beijing explored the history and philosophy of traditional ink painting in China and examined its relationship to contemporary art in China and the US.

Among other things, we discussed the difference between the brush stroke in Chinese ink painting and American

Abstract Expressionism, as well as the distinction between the western mark and the eastern stroke. This came to the fore in discussing the difference between ink as a type of stain—as a footprint, a residue, a trace, or what I call a "material index"—as opposed to the stroke, which pulls very distinctly from disordered phenomena. In ink painting, the brush stroke seems to capture the inner line of things that have no fixed form—the rocks, clouds, plum blossoms that vary. There are very specific strokes for each. But these are still same type held in mind by the artist.

The stroke in Chinese painting functions in many ways that would seem quite paradoxical to us in the US: it is both form and hue; the brush is both instrument and gesture; its stroke is both volume and rhythm; it is not a line without depth, like we would consider an outline, nor is it simply following the silhouette form. Rather, it is an animate stroke, and in this sense, it resolves the conflict between the representation of volume and the representation of movement.

Ink intimates hidden relations between concrete things and embodied gestures, drawing a coherent path in the stroke from the apparent disorder of phenomena. So, "to ink" is to follow the inner lines of things, to delineate an external reality.

The Rorschach Paintings Series leaves many meanings left open to the viewer. They are nonorientable, or multiple-ly orientable—providing different readings depending on which way you view them. They use many of the formal determinants of the Rorschach test, which is interesting because it is not a psychometric test. It has been developed over time to have a greater degree of scientific measure through statistical probability, and in looking at how doctors read the [Rorschach] score cards, I found a lot of interesting things to mine for part-whole relations in architecture.

The Projective Drawing Series plays with space to imagine new ways of visioning concrete in architecture. The field is notoriously bad with representations of fluid material—making symbolic hatch marks that miss much of the material potential of concrete. This series uses the fluidity of ink as an analog to the fluidity of concrete, using literal acts in ink like pouring or etching. This involves moving from tone to contour, and imagining a future act of construction, as well as the materials and methods that might allow slippages in the paintings to be enacted in material.

I started to think about not enacting the works myself. So the last series of works, *Concrete Poetry,* starts from what I like to call the "latent poetry of specifications" to describe an action in concrete, and to then define it through specifications. The concrete panels that make up this series are meant to be cast off-site in a factory, and to use this rational process to allow a very indeterminate result—one that indexes the body of the laborer and the circumstances of the site.

Mark Wigley

I love the perversity of this research. In architecture schools, we're dealing with the first post-ink generation. Of course there's a lot of ink in architecture, but it's hidden. Ink is inside a cartridge, and the cartridge is put inside a printer, and the moment that the ink makes contact with the paper is absolutely concealed. This is very similar to a general fear of liquids—usually a sexual

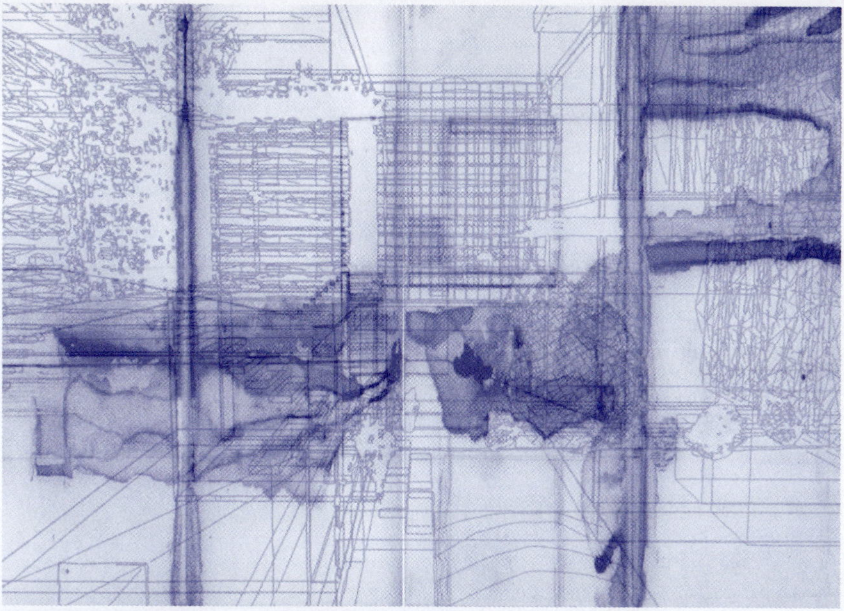

Concrete Poetry: Conceptual Acts of Architecture in Concrete

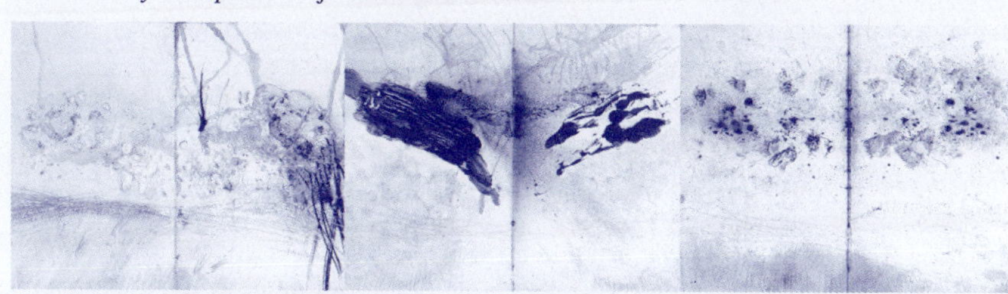

Rorschach Paintings Series

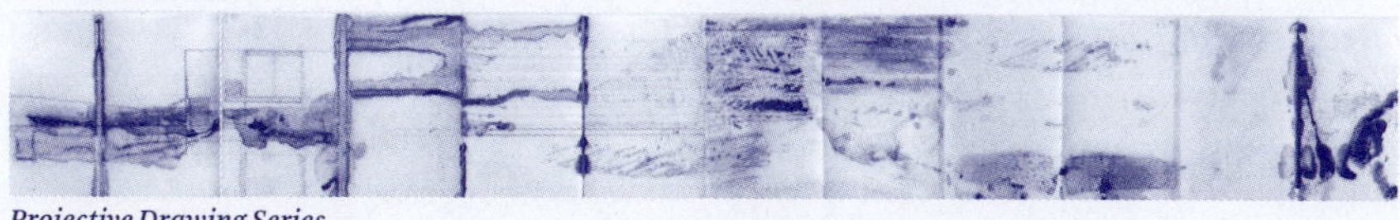

Projective Drawing Series

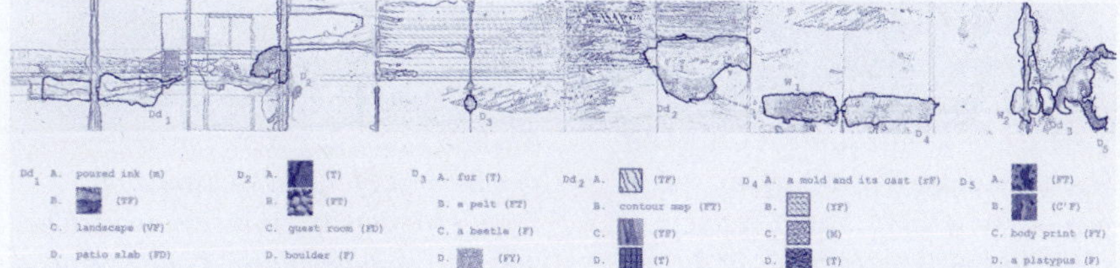

Multiple Choice Rorschach Test

fear. There's a phobia in architecture that [says] we should not see the release of the ink to the paper. Then ink appears not to be ink. The ink forms an image. Even if one has a sophisticated knowledge of which inks and which kinds of paper give the best image, it's read as an image, not as ink.

So considering this post-ink generation, for Michelle to launch an ink investigation is amazing. And it seems perfectly natural that she go to China to do it, because ink is China.

In architecture, ink is of course identified with the line—not just the mark, but the line. That means precision, clarification, specification, control, giving instructions. And it cannot be just any ink. Le Corbusier did not win the League of Nations competition because he used printers ink rather than China ink, but the rules of the competition stipulated China ink.

His being fired from that competition lead to the formation of CIAM— there was a decision made that they needed an international union of architects to control, to regulate the flow of commissions in architecture, and the flow of ideas. You can literally say that the mafia of modern architecture is premised on ink. You can further argue that the formation of the architectural profession is directly related to China ink. In the United States, for example, the profession started around 1830, and China ink was required—the crucial thing about it as distinct from printing ink or writing ink is that it is carbon. It comes from soot on a stick. So there's a very particular quality to making the line in carbon.

This black line—this super-black line—is usually done on tracing paper or vellum. The line is floating in space. And it's in the space between ideas and the physical world: a set of instructions that float in a hypothetical space of instruction, somewhere between the brain and the body, between ideas and the work. For this reason, to design something is a mental act, not a physical one. The power of the line in architecture is that it is a very, very precise definition both of the idea and of the world, and it's reversible. You can make a drawing of the world, and in so doing catch an idea. Or you can have an idea and project it through the drawing onto the world.

This exhibition explores the opposite logic: ink as a deep, dark pool, as a place of mystery. You are interested in the dip into the ink, and the stroke— these two gestures. Of course, the Rorschach is the exact opposite of the line. Not just because it's not a line, but because it's meant to be filled. The Rorschach allows you to see what your brain cannot see. So it's exactly not the drawing of the line.

The great thing about all ink, but China ink in particular, is that it's completely filled with different materials: the monolithic effect of ink is produced by animal, vegetable, and mineral materials in different combinations. The illusion of an infinitely monolithic material—which would allow any idea to be registered—is actually made possible by a chemical mix of organic and inorganic materials. In that sense, ink is filled with secrets. So it makes sense that you allow the mysterious nature of ink to be the main event.

Ink is always carries imminent catastrophe. Your shirt covered with ink. The page covered with ink. Your drawing destroyed at the last second by a mistake. So this whole logic that you treat as a kind of scientific research of the blot, of the leak, of the sphere and of the stain—this is also a trauma. This is like the disaster that ink always carries with it.

A big part of the discourse of ink is to hold it at the threshold, at the point of absolute disaster, to release it under control, and in so doing demonstrate your superiority over the world of liquids. One could even say that the heroic figure of the architect—controlling the world by controlling the line—is really about nothing more than controlling the ink.

Your research returns us to this release of the ink and all the secrets the ink spilled out. But instead of telling us that you made a mess, or that we made a stain, you're saying the opposite. You're making very precise research.

Qing Pan
[via simulcast from Studio-X Beijing]
Mark mentioned that Michelle came to China to undertake this project. I think that her work provides a lot of inspiration for Chinese artists working in ink. While they try to innovate, they do so under the weight of a long tradition of past innovations. Michelle's work comes from a very different perspective and training, and I think that gives a lot of insights to the artists.

I've worked with ink for a very long time. I've curated shows of contemporary ink artists—we had an exhibition of fifty artists, and a symposium of Chinese ink painters—and nobody's work is like hers. It is really unique.

Her work shows a lot of similarities with Chinese ink paintings. For example, there is attention to the negative space. The marks and the space around them are equally important. In Chinese we say, "from the white space we know where we place the black."

In traditional Chinese ink paintings, we have a character that means "to have a bamboo fully formed in one's mind" before one puts a mark onto paper. This term is so well-known that it is used to refer to anything that is fully prepared. Throughout the process of ink painting, the ink will have accidental effects on rice paper, and that leads to a variation of the first intention. But a bamboo—the intentions of a bamboo—will not transform into a blossom in the end. There is an intentional quality to the beginning that actually carries out to the end.

I find this different from Michelle's work. Hers is sparked from something that's already accidental, from some accidental image. Then, the intention is drawn out of this accidental image through her own imagination—and probably through your imagination—to become something very concrete.

Jonas Mekas
Thank you for this very enlightening, amazing lecture on ink. It's amazing. My background is not architecture. See, in the 60s and 70s—the film community—we used to be very much together in our meetings. Poets, musicians, dancers, but there were no architects. We did not learn much from them, and they did not learn much from us. I think it unfortunate, because it would have improved New York architecture. [Audience laughter]

As ink goes, my teacher in architecture is Raimund Abraham. Whenever I see him, he says: "Oh, architecture is going down. Whenever I walk into the big offices of the best architects today, all I see is computers, computers, computers. They never touch pens, pencils, nothing. It's only computers."

So now, here, I hear the some movements—a movement—towards ink. I hope it's real.

Pour/Tilt

Define volume/weight:

area of panel   = (1)A0 = 1189mm x 841mm = 1m^2
                       = (12)A0 = 12m2

thickness   = .15mm concrete layer, Nirvana pre-cast concrete panel
                .25mm maximum variance for structural

volume, panel   =(12)A0 = 12m2
                =(12m2) x (.15m thick) = 1.8m3

weight, concrete   = 2400kg/m3 Nirvana system
                = 1200kg/m3 Litecrete

weight, panel   = 4.5Mtons, Nirvana system
                = 2.25Mtons, Litecrete

weight, crane   = 6.4mtons max for pick/carry cranes on slope
                -20% allowance for safety required by code
                = 5.12Mtons

Define viscosity:

slope, site   = rise/run = 1m in 1m2 (841mm x 1189mm, 1456mm diagonal)
                = 1m/1.456m =.6868% slope
                = rise/run =.5m in 1m2
                =.5m/1.456m = .3434% slope

slump, concrete   = 75-100mm to pour on slope
                = 3-6mm per 300mm

slope, slump   = 6mm/300mm= .02% (same as required for drainage)

ratio of slopes   = site slope - slump slope / site slope
                = .94% difference to orthogonal
                = 1.06% difference to slump surface

Define resistance
Define vibration
Define pour sequence

Tilt Series and *Pour Series*

This is like *Urschleim,* the original mud from which something comes. You cannot put your finger on exactly how it works, but it's there. You *know* it's there.

What the pen does, and what ink does—the color, the stroke, all the amazing things that you enumerated—you can find that in a different way in every art. You can talk about the stroke in music, in painting, in dance. It's in all of the arts. Sometimes we don't know how to find it, or how to see it. But all of the elements that we find in ink—different aspects, colors, movements—can be talked about in any other art.

But, this is new. To me, this is beautiful.

Economics For The Rest Of Us

Moshe Adler, David Cay Johnston

February 11, 2010

Co-sponsored by the Harry van Arsdale Center for Labor Studies at SUNY Empire State College and The New Press

What good is Wall Street? How do exorbitant CEO bonuses affect all New Yorkers, as well as the physical structure and social life of their city? This book launch for Columbia GSAPP Urban Planning professor Moshe Adler brought to Studio-X New York a conversation on governance and bias in the field of economics, featuring Adler and Pulitzer Prize-winning journalist David Cay Johnston.

—

David Cay Johnston
The conversation about redistribution is always discussed in terms of taking from the rich, presumably the productive class, and giving to the poor, often characterized, subtlety perhaps, as the undeserving or unproductive class. But is it not possible that what we're really doing is redistributing upwards? That the government, by its rules, is causing people at the bottom to have less and people at the top to have more?

Moshe Adler
You hit the nail on the head, but I think that before we have redistribution we need justice in the division of what we produce together. We need simple laws: a limit on the ratio between shareholders' income and executive income on the one hand, and between the highest

and lowest paid employee within the corporation on the other. The latter is now on average about 430 and increasing over time with no relationship to profits. You can make tons of money while you are losing money for your corporation.

Executives have it easy. When it comes to workers, they say, well, we cannot raise your wage, because you are getting paid what you are worth. When it comes to my wage as an executive— my salary of $10 million or $100 million —well don't tell me that it is my fault that my corporation lost money, because we don't know what it is that each individual person produces. So, when it comes to the wages of workers, we cannot pay you any more because you get exactly what you produce. You cannot pay me any less, however, because how do you know what it is that I have produced? So we are losers at both ends, and we cannot solve it on our own. This is why we have a government. It is a government of the people. We have to realize that *we* are the people.

But my concern is really with the field of economics. It has been taken over by the rich. Every theory that made sense was empirically true and sided with workers and the poor was thrown out for absolutely no reason whatsoever, and replaced by theories that have no validity to them—particularly the theory of wages.

DCJ

I did a calculation a while ago about executive pay of senior executives over the last forty years, by which I mean the five main officers whose pay was disclosed. If the stock market grew at the same rate as their pay, the Dow, which has been hovering for a long time around 10,000, would be at 80,000. How do we get away from the economic equivalent of geocentrism?

MA

I'm not sure that I have an answer, but the fact of the matter is that the Great Depression gave us an opportunity to have Keynesianism, and that led our economic policy for 20–25 years— between 1935 and 1960 or so. The attack on Keynesianism started once the memory of the Great Depression faded. It was led by Milton Friedman and his number one water carrier, Ben Bernanke. They said that Keynes was wrong, and that had we left it to the market, everything would have been okay: the Great Depression was the fault of the Federal Reserve. At Friedman's 90th birthday party, Ben Bernanke gave a toast, saying, "Milton, you were right, we were wrong." The "we" was the Fed, although Bernanke was not the chairman at the time. He was basically saying "We are not going to let this happen again." And when it happened on his watch—this recession, this disaster we're in—he gave a trillion dollars to the banks.

Keynes would have found that laughable. He had said that periodically, for reasons that nobody can know, predict, understand, or fix—in particular not the government—investors lose their confidence. They lose their optimism, and this optimism cannot be revived. The government should not try to revive it, but should hire workers itself. FDR hired three million people to build the national parks, to paint and draw murals on public buildings, and he hired historians to do oral history: odd jobs that we needed, and that we still need. He knew that the government could not revive investment artificially, because people lose confidence and

do not recognize good investments during or after the Depression.

We see it right now. The banks are
not lending money, and very few people
are asking for loans. And why should
they lend out money? It would be
imprudent to lend money now, because
we don't recognize good investments.
Investors are also prudent by not invest-
ing, because again, they do not recognize good investments. So, giving
this money to the banks would not have
made a difference, and in fact it has
not made a difference. It did not
produce jobs. It only took a trillion dollars away that banks now claim they
don't have for the rest of us.

DCJ
I think what you're saying is that there
is a fundamental bias in neoclassical
economics. It is biased in favor of the
already-haves. All of us, including
the rich, would be better off in the long
run if we were to get away from this
inherent bias and think more openly
about what's happening in our society.

Ink Workshop with Karen Finley and Michelle Fornabai

Gabriella Bass, Brittany Beyer, Tracy Brown, Gavin Browning, Mary Calvani, Marjuan Canady, Jordan Carver, Esther Chang, Claire Dub, Karen Finley, Michelle Fornabai, Noelle Ghoussaini, Francisco Gomez Duran, Dalia Hamati, Richard Hamilton, Dorothy Jiji, Victoria Lynford, Kate MacGregor, Daisy Nam, Christina Nguyen, Lauren Nixon, Hye Lee Oh, Juan Ortiz, Vanessa Ramalho, Isabelle Rijnties, Shea Sabino, Clara Ines Schuhmacher, Reed Simonds, Andrew Vann, Wan-Jung Wei, Rodrigo Zamora

February 16, 2010

Ink is universal—but how is it understood across disciplines? Concurrent with the exhibition _ink_, architect Michelle Fornabai (Columbia GSAPP) and visual and performance artist Karen Finley (NYU-Tisch) brought their classes to Studio-X New York for a morning workshop and conversation on the various uses of ink in their work. Finley's textual and performance piece _The Passion of Terri Schiavo_ used ink drawings to comment upon the 2005 euthanasia controversy alongside the popularity of the film _The Passion of the Christ_. Finley's presentation was followed by a conversation on ink, Rorschach imagery and trauma. The students were asked to respond to the then-recent earthquake in Haiti, and were grouped together to create exquisite-corpse drawings using sumi ink and vellum.
___

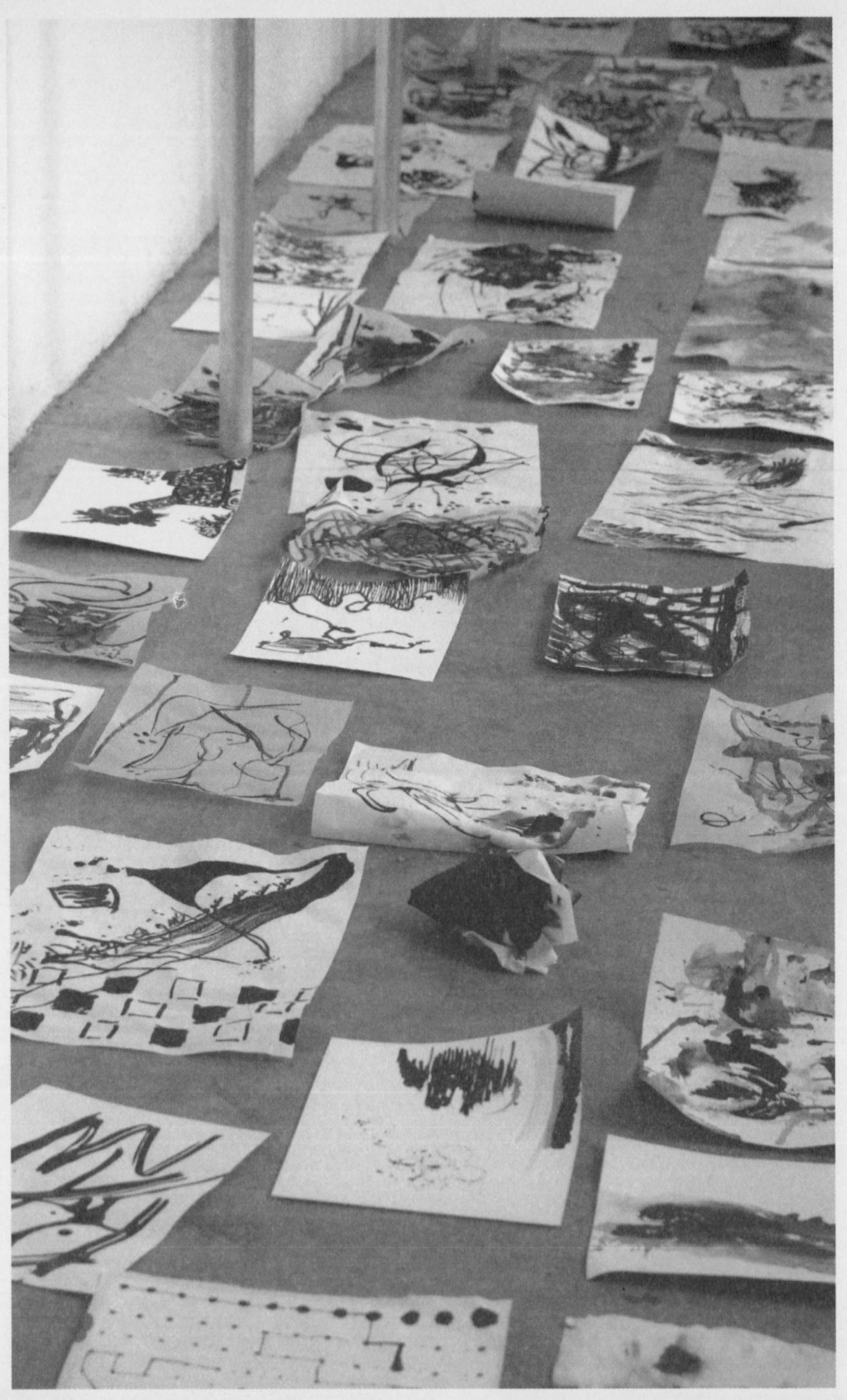

*Instead of hearing Muzak when
you walk into a building,
what if you heard the processes of
how that building was built?*

—Daniel Perlin, 03.18.10

*That it is so hard to imagine today
how young radicals in the 60s and 70s
could have believed that
their explosive 'communications'
with the government would bring
about real change shows how
profoundly the culture has shifted.*

—Sierra Feldner-Shaw, 11.17.08

*Why can't a table eat itself?
Tables support food,
but why can't they also be food?*

—Ate Atema, 05.14.09

*Kids don't need to learn
how to boil an egg. They need to learn
how to raise snails, which are
a much more useful urban meat.*

—Natalie Jeremijenko, 02.27.10

*One thing the memoir destroyed
for me is the romance
of a politically radical lifestyle.
But I won't miss it.*

—Minna Ninova, 11.10.08

*Computing is no longer something
we encounter on the screen,
but something that dissolves into
the environment.*

—Mark Shepard, 01.08.09

Tobias Armborst, Daniel D'Oca,
Mathan Ratinam, Georgeen Theodore,
Andrea Zalewski

February 16, 2010

Tobias Armborst, Daniel D'Oca and
Georgeen Theodore, sub-curators of the
2009 International Architecture Biennale Rotterdam, discussed their selec-
tions from the US for the international
event. Then, Mathan Ratinam and
Andrea Zalewski of the Moving Image
Lab at Columbia (MILC) screened *Cities
of Preferences,* a short documentary
about neighborhood segregation in
New York City. The film was produced
(with Toni Schade) during the previous
summer at Studio-X New York for its
premiere at the IABR.

——

It has been Interboro's distinct pleasure
to research community and the open
city for the 2009 International Architecture Biennale Rotterdam. As we tip the
rural/urban scale towards the urban
for the first time in history, it is a good
time to test the ideal of the open
city against urban phenomena (squat-
ter settlements, sprinkler cities, exurbs)
that Christopher Alexander and Jane
Jacobs could hardly have anticipated
half a century ago. Our research focuses
on America, which is now more than
60% suburban, and where new master-
planned communities continue to
sort Americans by narrowly-defined life-
styles. According to Bill Bishop, whose
book *The Big Sort* was recently endorsed
by Bill Clinton, America is "becoming
increasingly crowded with people who
live, think, and vote as we do," and has
become "so polarized, so ideologically

inbred, that people don't know and can't
understand those who live just a few
miles away." What hope do accessibility,
tolerance, and diversity have in such
a climate?

Our research suggests that while
such large-scale suburbanization does
indeed pose challenges to the ideal
of the open city, it also offers opportunities. Jane Jacobs celebrated "generators of diversity" might not have much
relevance in an environment where
mixed primary uses, small blocks,
aged buildings, and concentration are
few and far between, but we wonder
whether there might be other, less
tried and true means of bringing about
more openness in the built environment. What if architects and planners
reframed the open city as something
simpler, lighter, more everyday? What
if, instead of striving to cook up the
open city with ancient recipes like the
"generators of diversity," architects and
planners strived to single out the open,
inclusive experiences that people
have in the course of their everyday lives,
and then thought up ways to multiply
and enrich those experiences? As Robert
Venturi said, "Learning from the existing
landscape is a way of being revolutionary for an architect." Could the open city
be subtly slipped into a suburban
commute? A trip to the supermarket?

These are the questions we asked.
Relying for the most part on de
Certeauian "tactics of the weak," our
ongoing selection of projects all co-opt
everyday phenomena—fences, pool
sheds, big box stores—in an attempt to
slip open city experiences into peoples'
everyday lives.

The purpose of our installation *The
Arsenal of Exclusion/Inclusion* is twofold.
On the one hand, we wish to use this
dictionary of 101 things that open or

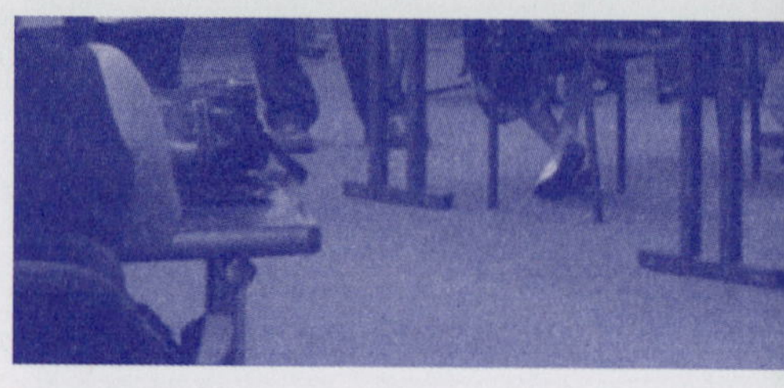

close the city to remind people that there are indeed good reasons to be critical of the suburbs. Especially in the 20th century, America developed an impressive array of institutions and policies that were successfully deployed to maintain spatial segregation, and that afforded anxious communities the ability to restrict access to those deemed undesirable. From overt "weapons" like Racial Zoning, Racial Covenants, and Racial Steering that have been used to restrict African Americans' access to emerging suburban housing markets in the 1950s, to more subtle ones like Conditions, Covenants, and Restrictions that are used today to maintain class-based suburban segregation, 20th century urbanization in America cannot be understood without examining these weapons of segregation.

We also wish to highlight tools that have been deployed to open suburbs: Forced Busing, Community Benefits Agreements, Community Development Banks, Good Neighbor Agreements, Home Value Insurance, Homebuying Workshop, Housing Court, Housing Vouchers, Inclusionary Zoning. Our ongoing research into these later weapons focuses on everyday phenomena. GPS navigation, Flat Fares, Halloween, Jury Duty, and Designated Smoking Zones are not typically thought of as things that open the city, but we demonstrate that architects would be wise to co-opt them.

—Interboro

---

t to make the church
Starbucks "third-
mall planning to house
der one roof. This new
interaction that go
core. These venues
community, ranging
work-out facilities,
s.

suburban locale,
ingly, it's no surprise
s is host to the largest
the one time Rockets'
using the new title of
ersed over five cam-
urches have become
ies of the centralized
the residents of the

public, these
an, formal, and pro-
drive-to enclaves
rnalized planning, they
r their size to a more

Size/

-4

found in munici-
size that a

---

building can be built
Arsenal of Exclusion
have used Minimum
ferences clear spatial
that are larger than th
metropolitan area, m
don't have the means

New Canaan, Connecticut v
Minimum Lot Sizes for this
Agreement: Discrimination
Kenneth Jackson writes ab
used in New York City in 19
minimum lot size, at the tin
New York City. The purpose
class of the town and to pro

— Interboro

See the following communi
Jumbolair

Model*

In an abstract sense,
inclusionary. Models
ties of a building or ur
fore be an important p
process, but models c

— James Rojas

*An interesting example of
Transportation Planner Ja
(DSUP), on view in the the
Way of Living."

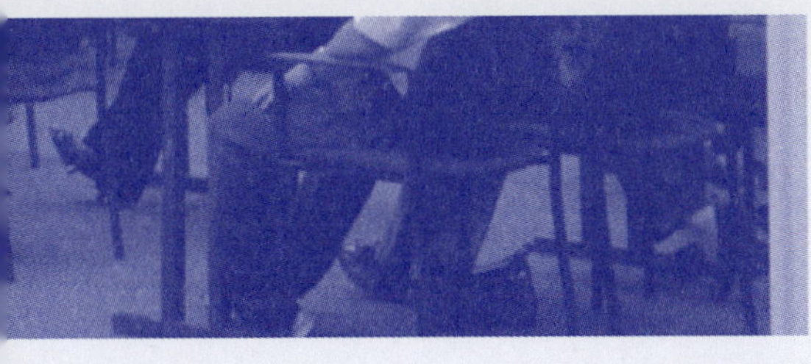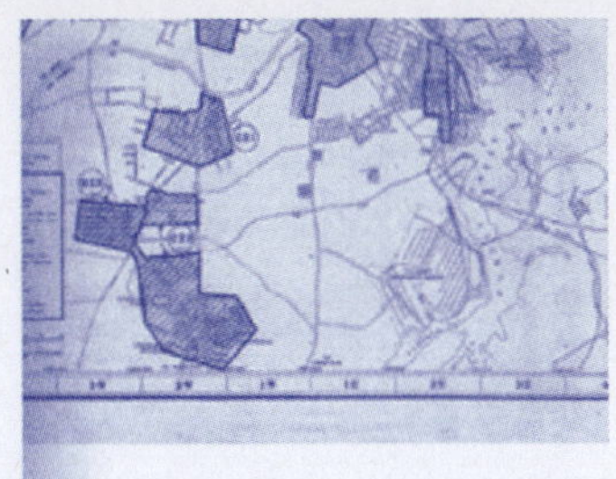

um Lot Size is in the
...historically, municipalities
...egulations to keep class dif-
...ablishing minimum lot sizes
...nd in other parts of their
...ies can exclude people who
...ase large lots.

...he first municipalities to use
...the essay "Gentlemen's
...litan America," the urban historian
...venteen years after zoning was first
...naan's higher-ups zoned a two-acre
...lly large lot for a town so close to
...e-lot zoning was to preserve the
...ty values.

...ed in the Mural: Cottesmore,

...s neither exclusionary nor
...inly elucidate the complexi-
...gn scheme, and can there-
...lemocratic decision-making
...educe and deceive.

...el can be used to open the city is
...s "design-based urban planning"
...nd of "Community: The American

Mortgage Discrimination ❶ ❷ M-

This tool is neither a discrete "thing" nor has it been lega
for decades, but nonetheless continues to exert a profound impact on residential patterns and inequality in the
United States. Its origins are most readily located in the
U.S. government's creation in 1934 of the Federal Housin
Administration (FHA), a policy response to the havoc
wrecked on the nation's housing economy by the Great
Depression. The decline in home sales and housing start
coupled with a spate of mortgage foreclosures, placed
the construction, real estate, and home finance sectors
on the verge of collapse. The Roosevelt administration
responded with the FHA, a regulatory and mortgage insu
ance program designed both to revive and stabilize these
industries and to create and sustain a new kind of marke
for homes.

The program insured institutional lenders who agreed to issue FHA-
approved mortgage loans — for home improvement or purchase — to
individual borrowers, promising to indemnify lenders in case of default.
Participating lenders were required to use a particular type of financial
instrument: the long-term, low-interest, fully-amortized mortgage pio-
neered in the 1920s but largely untested by most financial institutions. T
new, FHA-approved mortgage had very low interest rates, required sma
down payments (as little as 10 percent), and allowed borrowers 20 (and
eventually up to 35) years to pay off both the principal and interest, tern
that made homeownership feasible for most middle-income Americans

The FHA's promotion of the new mortgage instrument quickly made it
a national standard, in both the government-insured and "conventional"
(or non-insured) markets for home finance. When combined in 1944 with
the Veterans Administration's mortgage "guarantee" program — which

Amale Andraos, Sean Basinski, Joel Berg, Jonathan Bogarín, Marcelo Coelho, Nevin Cohen, Makalé Faber Cullen, Rebecca Federman, Stanley Fleishman, William Grimes, David Haskell, Natalie Jeremijenko, Naa Oyo A. Kwate, Annie Huack-Lawson, David Sax, Beverly Tepper. Moderated by Geoff Manaugh, Sarah Rich, Nicola Twilley.

February 27, 2010

Well over 250 people came to Studio-X New York for four panels over the course of one Saturday afternoon. Panelists discussed food and the city—from transportation systems and bodega supply chains to flavor science and the spatial distribution of Jewish delis. But the real star of the day was the surging crowd. Every imaginable space was filled with people: seated in chairs, on tables or the floor, leaning precariously against windows and the artwork on the walls. Panelists struggled to reach the front of the space, and the bookseller complained that no one could see the copies of *Appetite City* or *Gastropolis* on display, let alone buy them. Yet, everyone marveled at the amount of public interest in food as a design tool. A few days later, the building's management issued a complaint.

—

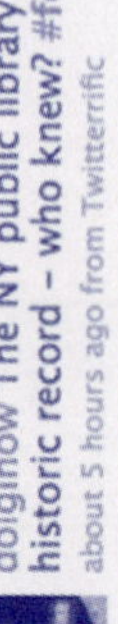

yeahlikethat Hipster douche insists on recording #Foodprint with his dinky p&s. Guess he doesn't know this is all being pro filmed and posted online.

about 5 hours ago from txt

dolginow The NY public library collects menus as part of it's historic record – who knew? #foodprint

about 5 hours ago from Twitterrific

parkview Wow, from a peak of between 2500 and 3000 jewish delis in the 5 boroughs, there are now about 25. #foodprint

about 5 hours ago from Tweed

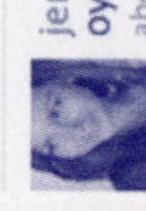

jenniferspaeth Pearl Street named after the abundance of oyster pearls chucked from the seaport #foodprint

about 5 hours ago from UberTwitter

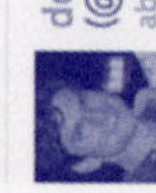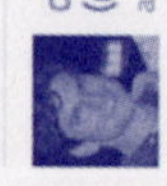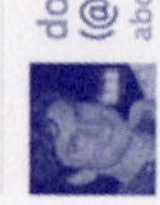

dolginow Excited to get my NYC food thoughts on #foodprint (@ Studio-x w/ 9 others) http://4sq.com/8XomFg

about 6 hours ago from foursquare

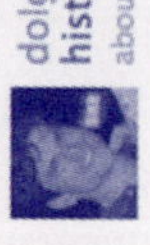

vercheesey RT @raganella7: If 10% of 52k acres of private yards in NYC grew veg = 113 million lbs of veg/yr to feed 700k NYers. Nevin Cohen #foodprint

about 3 hours ago from Twitterrific

BklynFoodCoaltn RT @SkeeterNYC: TODAY – **FoodPrint NYC** @foodprintcity. Free. Check out the amazing panelists and **panels:** http://www.foodprintproject.com/

about 3 hours ago from TweetDeck

NYFarmer @parkview Why do u think there was anti Pollan undercurrent @ **Foodprint?** just askin'

about 3 hours ago from mobile web

curtisfinancial This look like a really cool event in NYC and I love the logo: http://bit.ly/9dEcOE **Foodprint NYC**

about 3 hours ago from TweetDeck

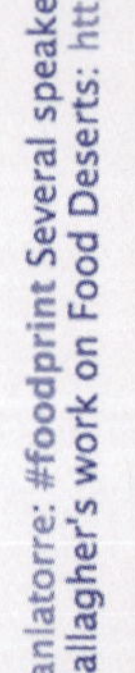

NYFarmer RT @danlatorre: #foodprint Several speakers extending Mari Gallagher's work on Food Deserts: http://bit.ly/aueOnR

about 3 hours ago from Tweed

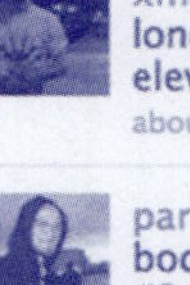 danlatorre #foodprint Beverly Tepper – we don't need more tech & tools, to get people to respect food teach them to cook.
about 4 hours ago from Echofon

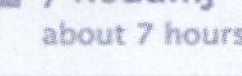 ChiefHDB RT @parkview: Wow, from a peak of between 2500 and 3000 jewish delis in the 5 boroughs, there are now about 25. #foodprint
about 4 hours ago from Tweed

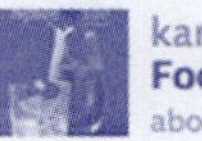 SkeeterNYC Interesting! RT @parkview Wow, from a peak of between 2500 and 3000 jewish delis in the 5 boroughs, there are now about 25. #foodprint
about 4 hours ago from web

 SkeeterNYC @dolginow There were no food guides in NYC until 1903 despite there having been a sophisticated food scene for 50 years. #foodprint
about 4 hours ago from web

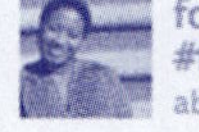 dolginow Really clever: plant carbon eating plants in front of no parking area around fire hydrants, but the city loses ticket revenue. #foodprint
about 4 hours ago from Twitterrific

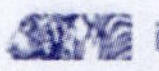 mikemakmyday At foodprint discussion. Varick st. Urban ag. Awesome http://bldgblog.blogspot.com/2010/01/foodprint-nyc.html
about 4 hours ago from TweetDeck

parkview First restaurant guide: "where and how to dine in NY" 1903 available at archive.org #foodprint
about 5 hours ago from Tweed

dolginow There were no food guides in NYC until 1903 despite there having been a sophisticated food scene for 50 years. #foodprint
about 5 hours ago from Twitterrific

 randyjhunt Pretty sure that's Michael Pollan listening in at Foodprint.
about 6 hours ago from Tweetie

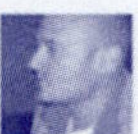 xmalikowskix #foodprint is packed! Should have gotten here long before it started. It's crowded as soon as you get off the elevator. Fearing for life.
about 6 hours ago from Echofon

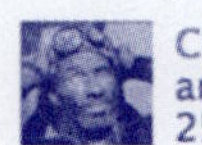 parkview Great question about the "informal trade" in bodegas. How do we map hidden sources of revenue? #foodprint
about 6 hours ago from Tweed

NYFarmer @fiftykeels Yay, keep livetweeting from FoodPrint, cool, we and cows listening on farm
about 7 hours ago from mobile web

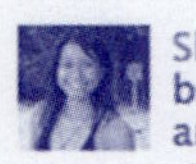 raganella7 Crazy turnout for #foodprint Space way too small. Shouldve known food would be a draw! http://yfrog.com/4fsudmj
about 7 hours ago from Twitterrific

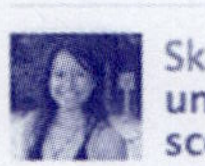 meredithmo Foodprint NYC and WOW it's packed. (@ Studio-x w/ 5 others) http://4sq.com/8XomFg
about 7 hours ago from foursquare

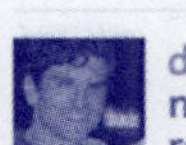 karanewman Working weekend. Bummed to be missing the Foodprint panels/exhibits, but duty calls...
about 7 hours ago from web

parkview Insanely over capacity @ foodprint nyc. :-(
about 7 hours ago from Tweed

foodculturist Can not hear one word @foodprintcity #foodprint
about 7 hours ago from Tweetie

 ifyoumustknow Awesome RT @raganella7 If 10%of 52k acres

TRANS SIBERIA

Warm Engine (Greta Hansen and Cheryl Wing-Zi Wong)

March 26–April 16, 2010

"The challenge? Trace the historic spread of communist ideology through building typology. How? Travel the 5,000 miles of the Trans-Siberian Railway from Moscow to Beijing during the dead of winter, stopping off at fourteen cities along the way. Get off at each stop to photograph and draw the administrative buildings and centers of power. Then, return to the United States, where you will quickly design and execute an exhibition from these findings."

Warm Engine departed Moscow on January 28 and arrived in Beijing on February 19, 2010.

Quote from Gavin Browning, "TRANS SIBERIA," *Places: Design Observer*, 3.26.10

—

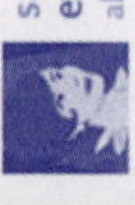
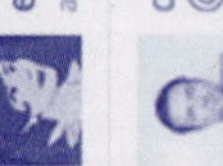
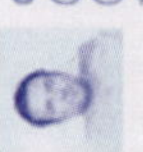
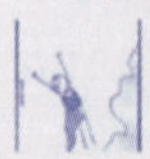

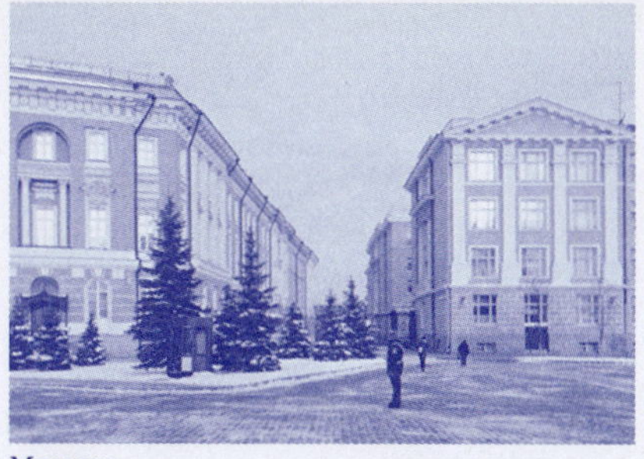

Moscow

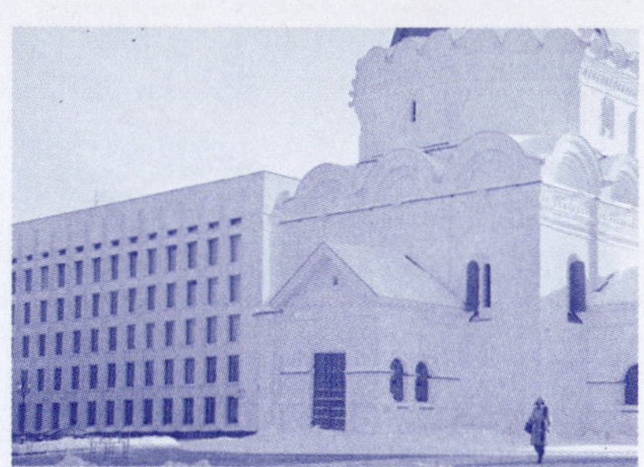

Nizhny Novgorod—461 km from Moscow

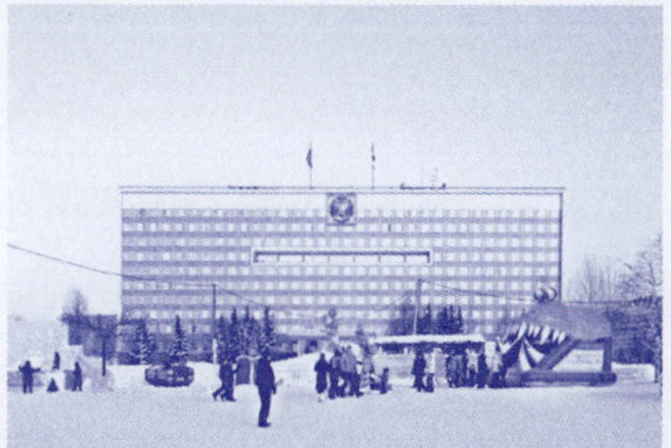

Perm—1397 km from Moscow

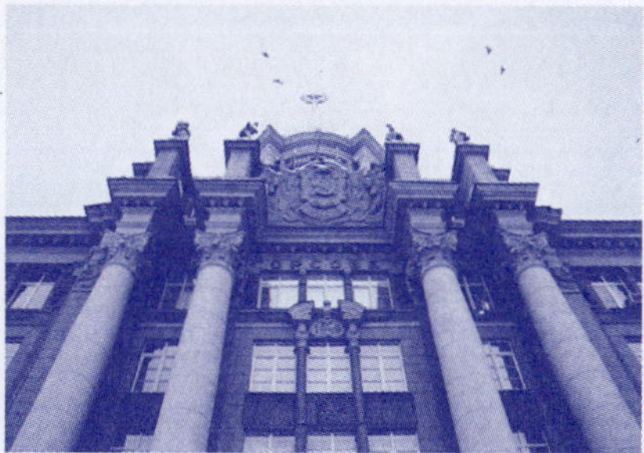

Yekateringburg—1778 km from Moscow

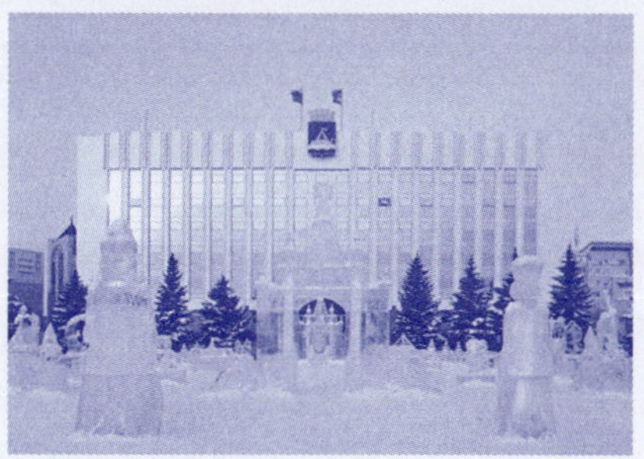

Tyumen—2104 km from Moscow

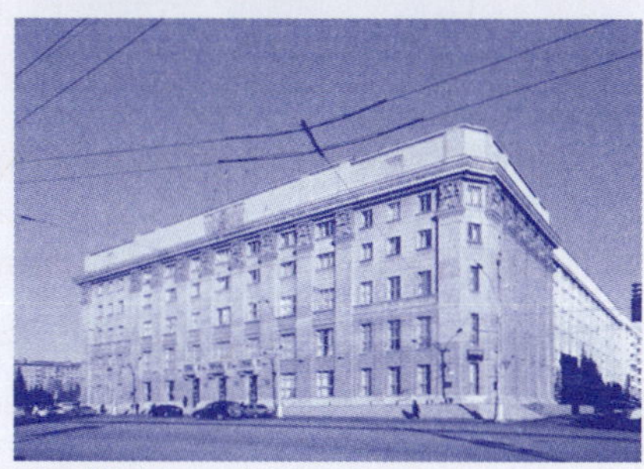

Novosibirsk—3303 km from Moscow

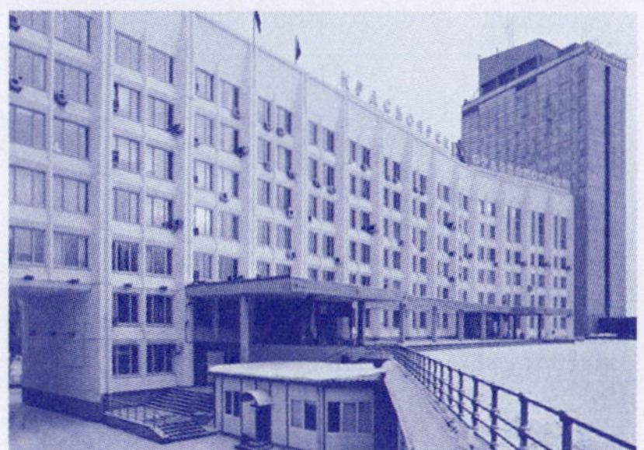

Krasnoyarsk—4065 km from Moscow

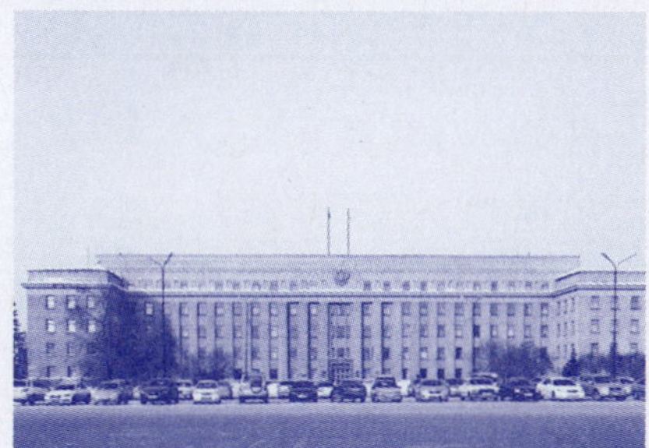

Irkutsk—5153 km from Moscow

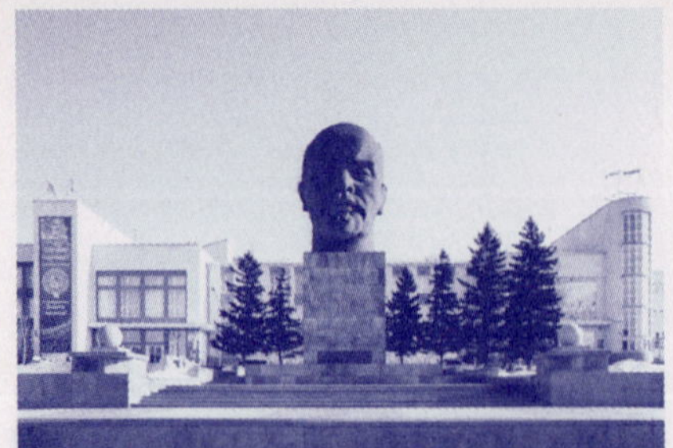

Ulan Ude—5609 km from Moscow

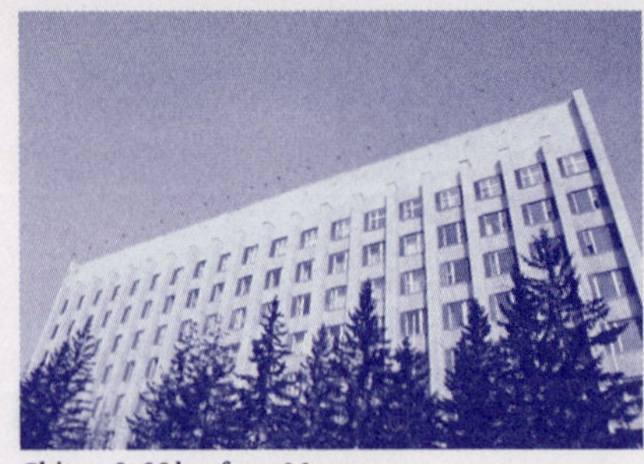

Chita—6166 km from Moscow

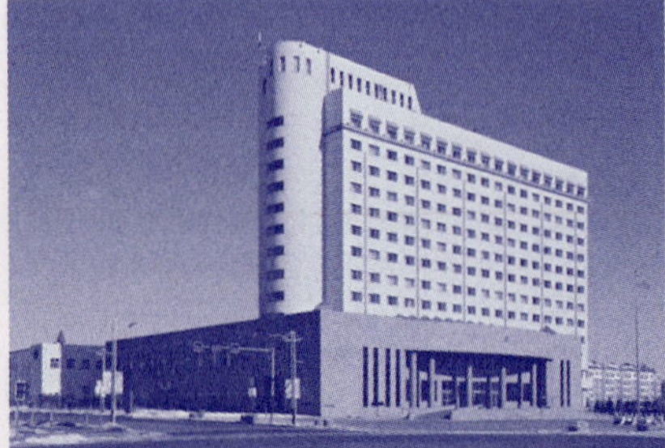

Manzhouli (China)—6638 km from Moscow

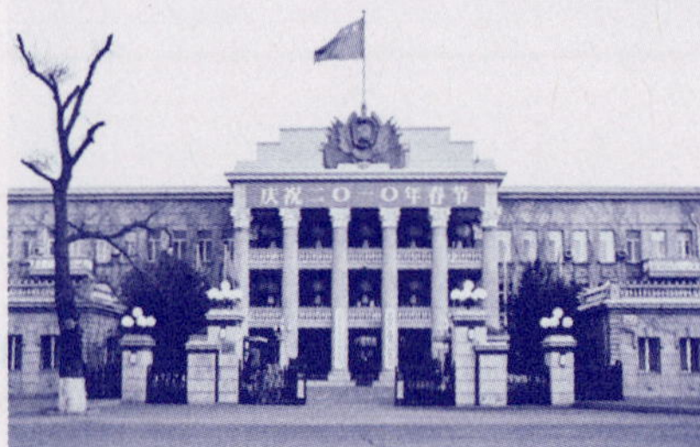

Harbin—7573 km from Moscow

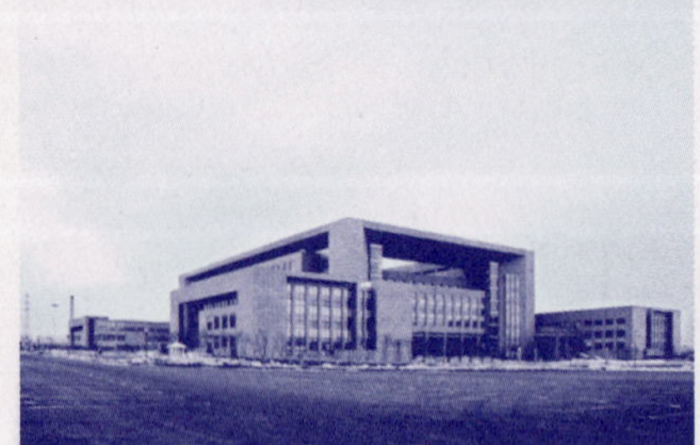

Changchun—7820 km from Moscow

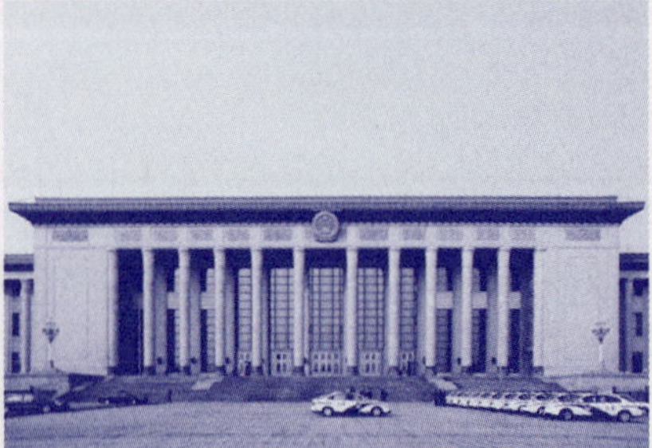

Beijing—8961 km from Moscow

*In the future, we'll continue
to be stunned when things go horribly,
massively wrong—at a scale
that we couldn't have predicted.*

—Molly Wright Steenson, 05.04.10

*The problem is not preserving
substance—transforming
the Narkomfin into a luxury hotel,
Melnikov's House into a museum,
workers' clubs into casinos—
because that has already happened.
The problem, I would say,
is the preservation of utopia,
the preservation of context,
the preservation of traces, memories
and programs of the earliest
Soviet Union: a society that
was aspiring to a radical change
in human history.*

—Jean-Louis Cohen, 12.02.08

*What moves the world is not tanks
and bombs, but human affect:
the aesthetics and sensibilities
of ordinary life,
and the drive to interact.*

—Harvey Molotch, 04.07.09

*I try to produce atmospheres—
spaces that have very specific qualities,
hopefully memorable spaces,
so that the notion of space and the
differentiation of spaces is taught.*

—Henry Urbach, 12.04.09

This is new. To me, this is beautiful.

—Jonas Mekas, 01.28.10

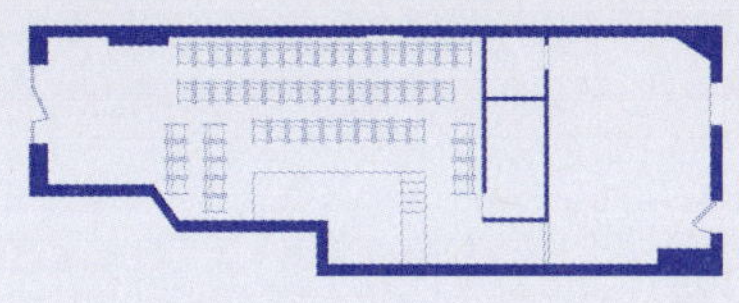

Jace Clayton (DJ /rupture), Mitch McEwen

April 1, 2010

In January 2008, Mitch McEwen opened SUPERFRONT, a gallery and event space in Bed-Stuy, Brooklyn, and shortly thereafter added a West Coast branch at Los Angeles's Pacific Design Center. Dedicated to architectural experimentation and interdisciplinary exchange, SUPERFRONT's wide range of overlapping uses melds design and community organizing, performance and political economy. At Studio-X New York, McEwen launched the first in a series of *Architecture Mixtapes*—created in collaboration with Jace Clayton (DJ /rupture)—and presented highlights from the LA show *UNPLANNED: Research and Experiments at the Urban Scale*.

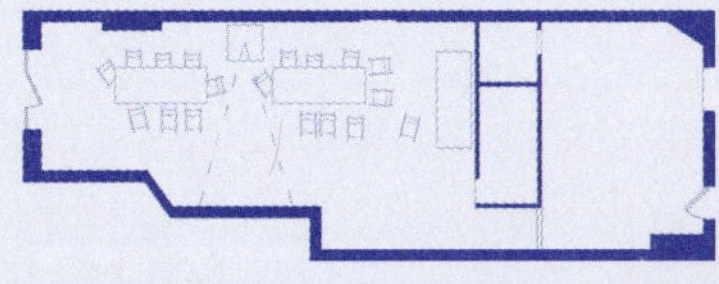

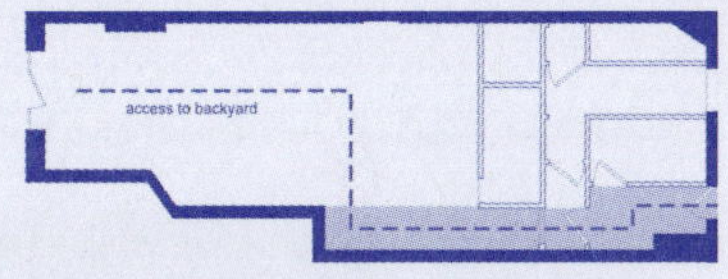

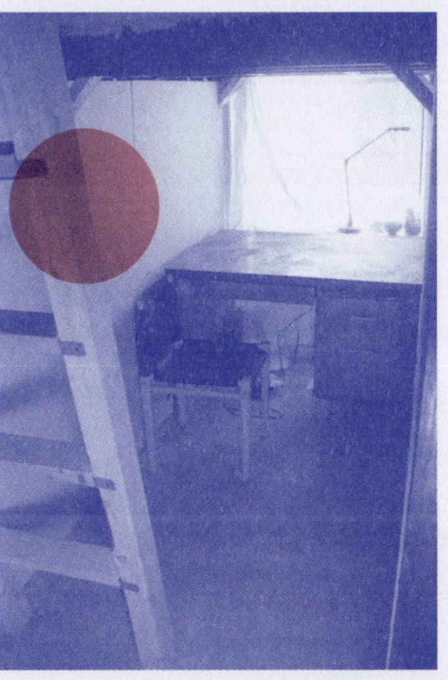

   McEwen also discussed one of the Brooklyn gallery's unique revenue streams, a benefit of the space's multi-use flexibility: two private yet interconnected living spaces. Hidden behind the public area, they may be rented on a short or long-term basis. The LA gallery has a different means of getting by: free rent.

—

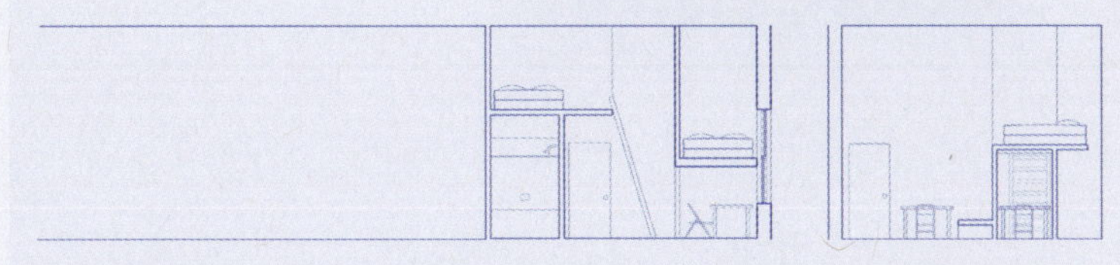

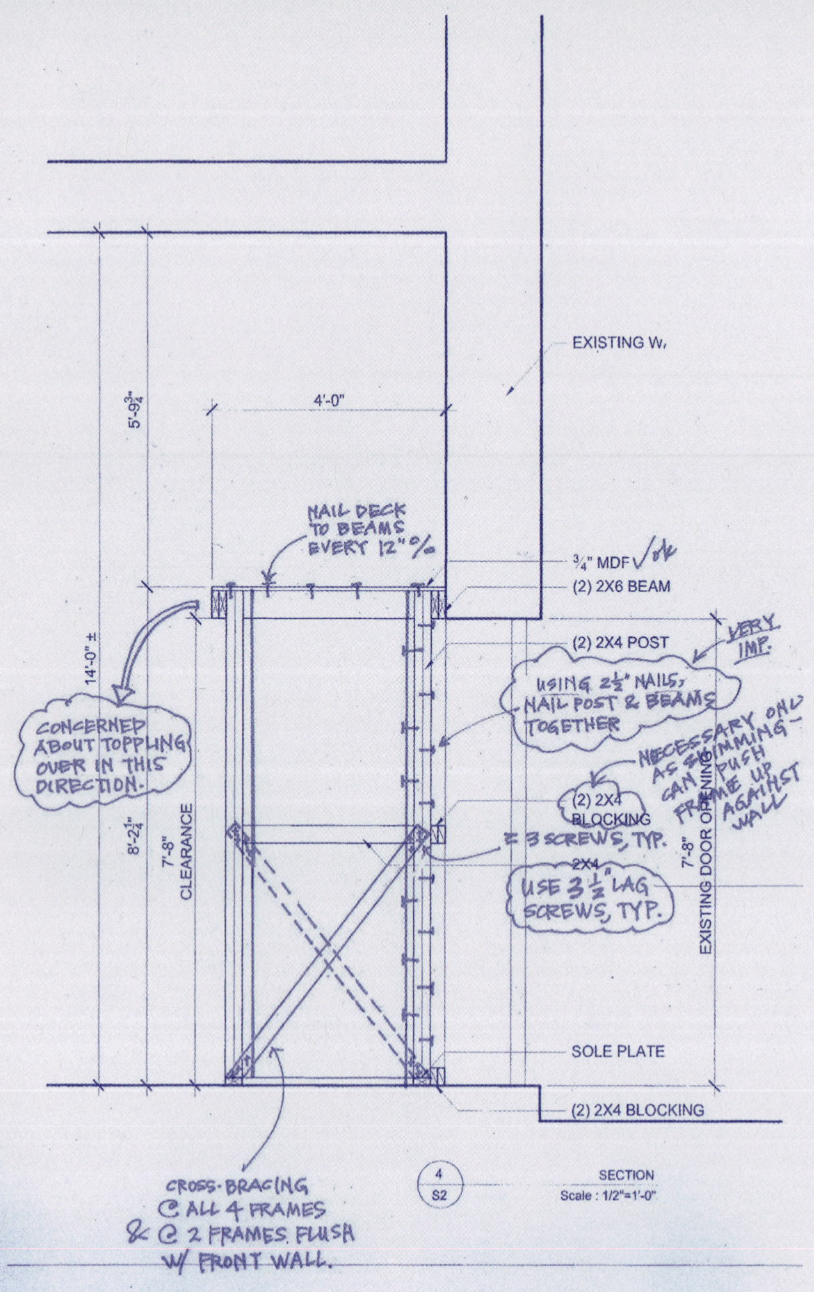

EXISTING W.
NAIL DECK TO BEAMS EVERY 12" °/c
3/4" MDF ✓ ok
(2) 2X6 BEAM
(2) 2X4 POST
USING 2½" NAILS, NAIL POST & BEAMS TOGETHER
VERY IMP.
NECESSARY ONLY AS SHIMMING — CAN PUSH FRAME UP AGAINST WALL
4'-0"
5'-9¾"
14'-0" ±
8'-2¼"
7'-8" CLEARANCE
CONCERNED ABOUT TOPPLING OVER IN THIS DIRECTION.
(2) 2X4 BLOCKING
≥ 3 SCREWS, TYP.
2X4
USE 3½" LAG SCREWS, TYP.
EXISTING DOOR OPENING
7'-8"
SOLE PLATE
(2) 2X4 BLOCKING
CROSS-BRACING @ ALL 4 FRAMES & @ 2 FRAMES FLUSH W/ FRONT WALL.
4
S2
SECTION
Scale : 1/2"=1'-0"

Imagination Vessels

Studio Sangue Bom, Sergio Cezar

April 22, 2010–May 6, 2010

The term "cultural exchange" is often heard today, but what does it mean? For Studio Sangue Bom, it means moving beyond language and geography, to physically manifest something that does not belong to any one particular culture. This is what prompted Keith Kaseman and I to orchestrate a one-week, experimental collaboration with our students and the Brazilian artist Sergio Cezar. The resulting *Imagination Vessels*—poised between New York City and Rio de Janeiro—are hybrids of Cezar's low-tech philosophy: "recycle the city: recycle your view." The spatial constructs were fueled by live exchange of ideas and materials, made in the United States with waste collected and carried from the streets of Brazil.

Exhibiting such hybrid work at Studio-X to a non-specialist audience allowed full realization of our exchange. The work was enhanced by the space's working atmosphere, different from the minimalist austerity of so many Chelsea galleries. There is actually a sink in the middle of the main exhibition wall! What could have been an impediment for some was an opportunity for us. Placing neon lights above the sink transformed it into another piece—not by hiding it, but through embrace and celebration.

—Raul Correa-Smith

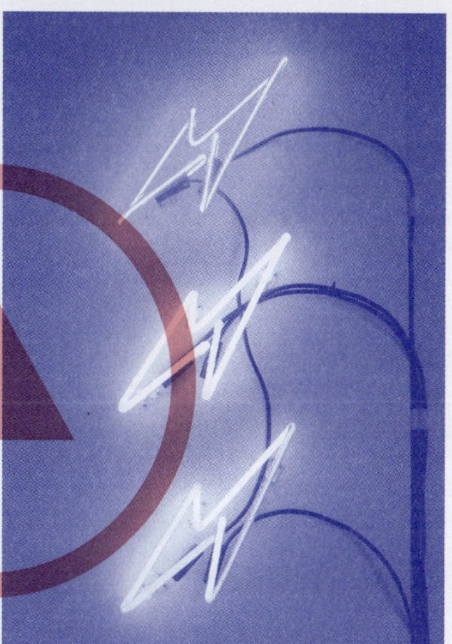

Studio Sangue Bom 2010: Keith Kaseman, Raul Correa-Smith, Steven Garcia; Yihong Deng, Hajar Ebrahim Darbandi, Keith Greenwald, Sara Gutierrez Armesto, Tzu-Husan Hsu, Analdur Schram, Lior Shlomo, Irene Urmeneta, Diego Urrego, Nicolas Weiss, Li Yang Wu, Daniel Yep

Sarah Cloonan, Pollyanna Rhee,
Federica Soletta, Tong Tong

May 13, 2010–June 15, 2010

Location: The Fashion Center Space
for Public Art, 215 West 38th Street,
New York City

Co-sponsored by CCCPArch and
The Fashion Center Business
Improvement District

A mixed-media, site-specific exhibition
about New York City's Garment District,
Dress Local provided a view into the urban composition of the neighborhood.
A handmade dress and silkscreened
maps—charting fabric, patterns and
notions—were portals into this unique
district. The exhibition's artifacts
were then displayed to the public in a
storefront window on West 38th Street,
comprising Studio-X New York's
first sustained street-level presence.

—

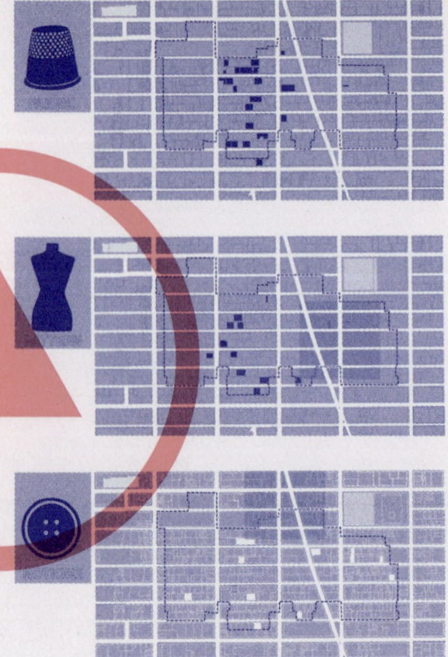

Moderated by Kazys Varnelis
Culture: February 9, 2010
Place: March 25, 2010
Politics: April 13, 2010
Infrastructure: May 4, 2010
Publish: June 24, 2010
——

During the last fifteen years, architecture and the media have been turned on their head as technologies of production and communication integrated into our daily lives. Instead of the delirious optimism of the last decade, we now also face panic and crisis. The media industry is in flux: as new media rise, old ones are victims of creative destruction. The tools of architectural production, meanwhile, have been thoroughly transformed; yet thanks to technological and legal innovations that made possible the securitization of buildings, architecture faces its greatest economic crisis since the Depression. If we can be certain of anything, it's that as Karl Marx wrote, "all that is solid melts into air."

In 2008, the Network Architecture Lab published *Networked Publics*, a book produced in collaboration with the University of Southern California's Annenberg Center for Communication that examines how social and cultural shifts centering around new technologies have transformed our relationships to (and definitions of) place, culture, politics, and infrastructure.

In spring 2010, the Netlab returned to explore the ramifications of these changes, giving particular attention to architecture and cities. In a set of four panels—culture, place, politics, and infrastructure—held at Studio-X New York, we discussed the consequences of networked publics in detail. Our goal was to come to an understanding of the changes in culture and society and how architects, designers, historians, and critics might work through this milieu. Discussions were broadcast live using uStream, and individuals worldwide submitted questions via Twitter.

Starting in the summer of 2010, the Netlab is collaborating with *Domusweb* on *Networked Publics: Publish*, a new collaborative publication. We invited brief submissions addressing the consequences of these changes for the architectural community. What transformations are taking place in the architectural profession, in architectural media, in criticism? How are these transformations interconnected? What do these mean to you? On June 24, an editorial board composed of panelists from the earlier discussions together with the Netlab staff met at Studio-X and reviewed the initial round of submissions.

In working with *Domus* at this key moment, the Netlab envisions that this publication will have various forms, both online and off, evolving and mutating over time. *Domus*, one of the earliest and historically most influential architecture magazines, sets itself as a case study for debate around the role of printed magazines in the contemporary era. If the magazine is no longer spontaneously embraced as a locus for debate, should the permanence of printed matter induce it to serve as a historical register for ideas developed elsewhere, e.g. on the Web (the magazine understood as an archive-in-progress of excellence)? Or, conversely, should it pursue agility, hybridizing across platforms? Does the notion of architectural criticism, understood in conventional terms, bear any relevance today? What forces designate the formal and conceptual frameworks of contemporary built architecture?

We are keenly aware that engaging with these epochal transformations will define the critical output of our generation. This publication is intended as forum for debate through which the accepted understanding of the word *publication* itself can be challenged, redefined, dismantled and rebuilt.

—Kazys Varnelis

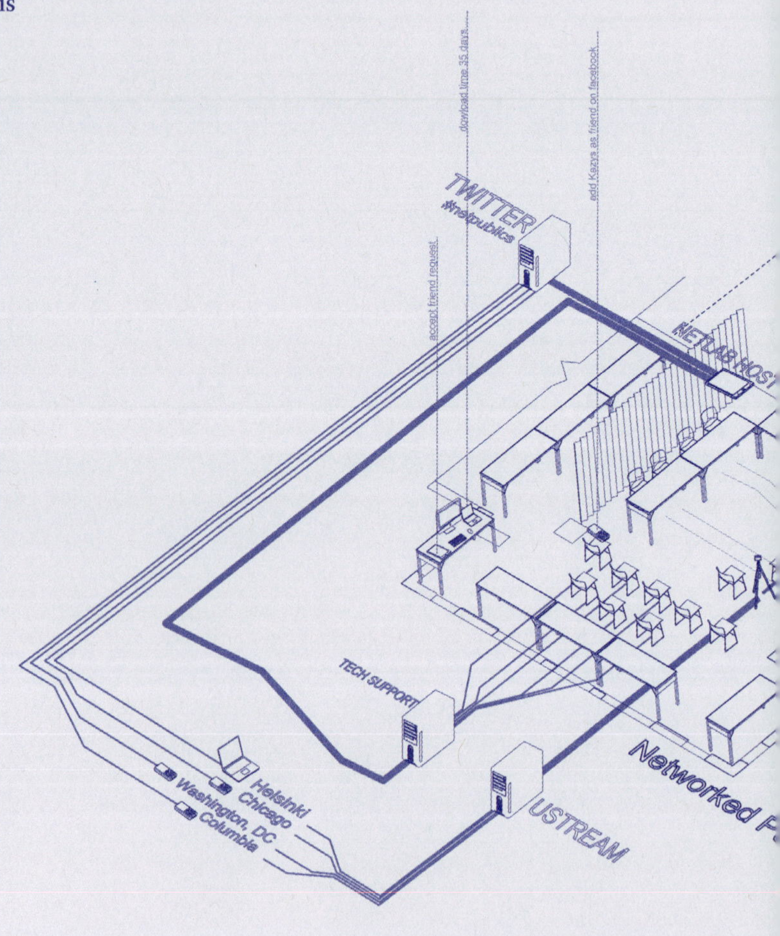

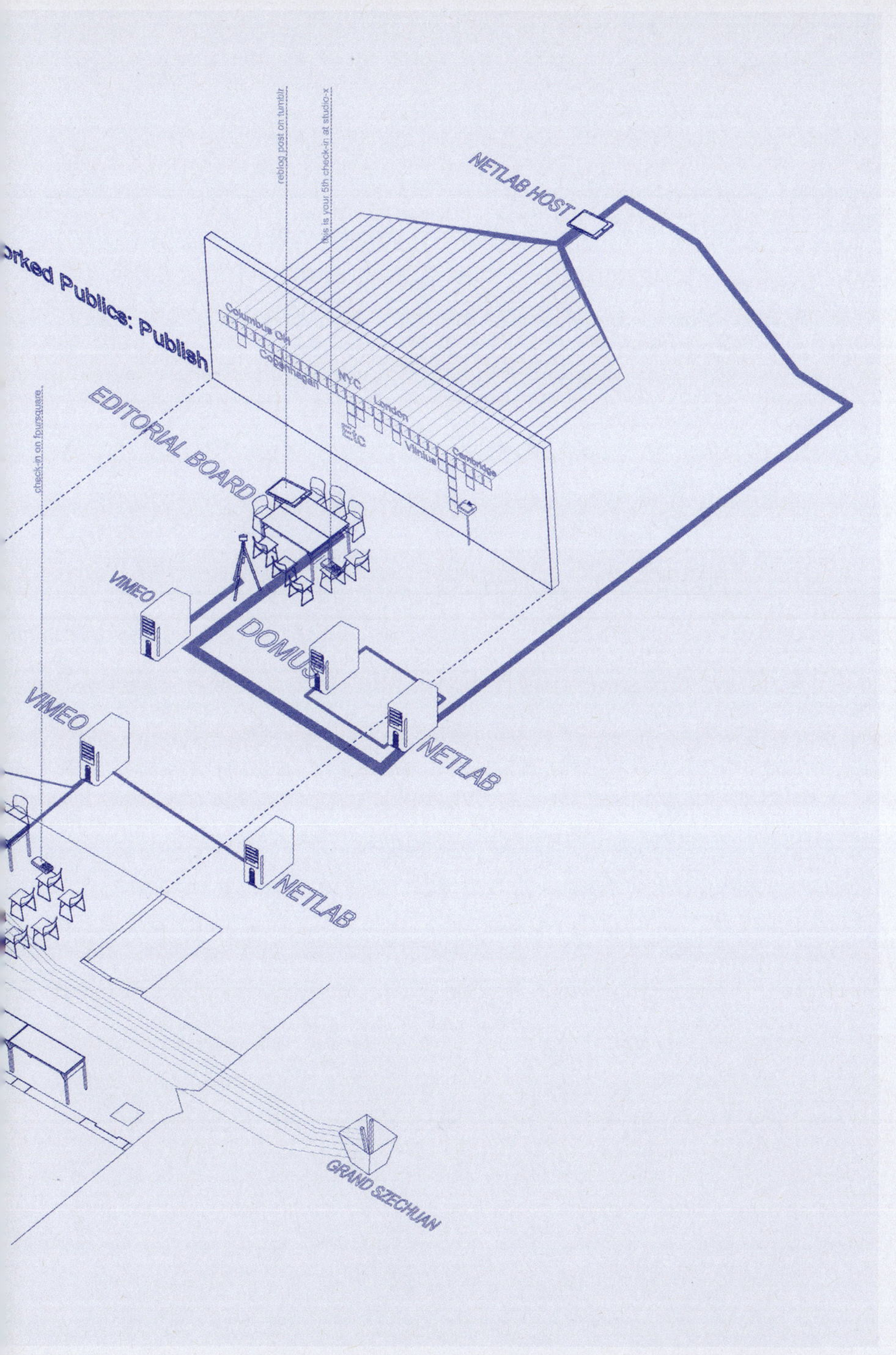
orked Publics: Publish
NETLAB HOST
reblog post on tumblr
this is your 5th check-in at studio-x
check-in on foursquare
EDITORIAL BOARD
Columbus Cin
Copenhagen
NYC
London
Etc
Vilnius
Cambridge
VIMEO
DOMUS
VIMEO
NETLAB
NETLAB
GRAND SZECHUAN

Rapid Response: Spontaneous Architecture

PRE-Office (Zachary Colbert, Aaron David, Daniel Kidd, Leah Meisterlin)

An ongoing monthly design competition based on current events.

——

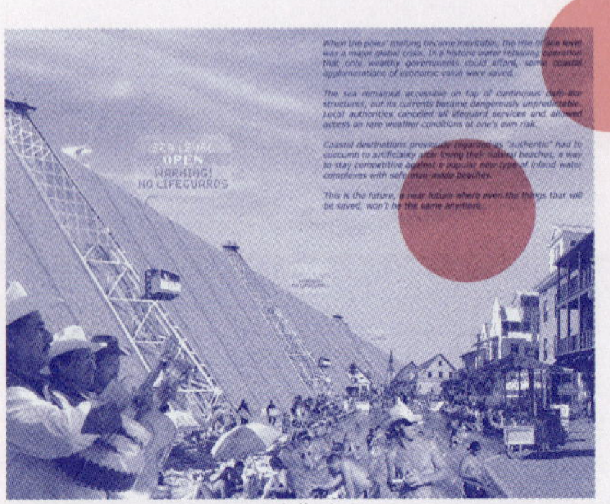

January 2010: The Future

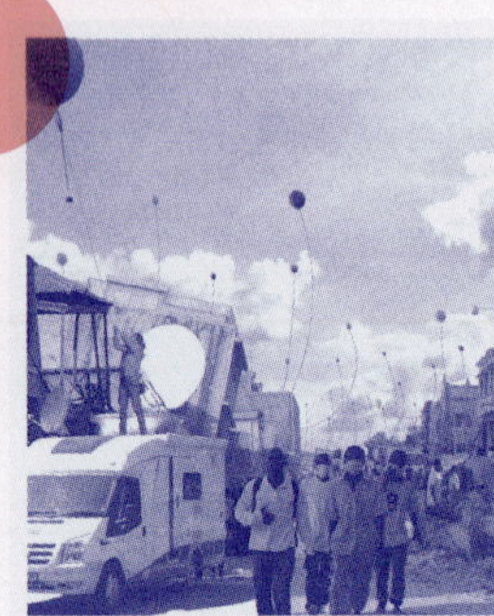
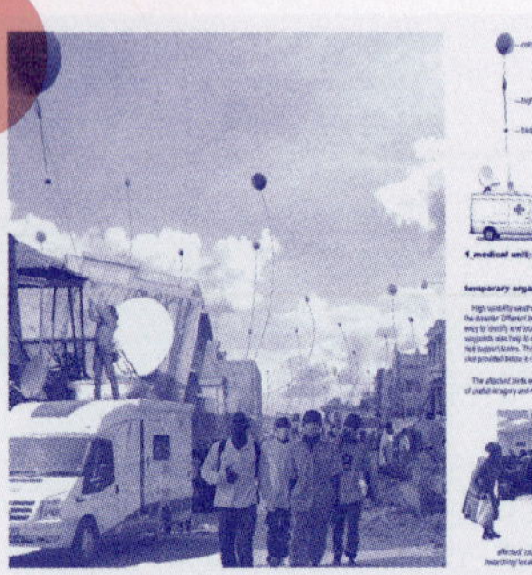

February 2010: Haiti

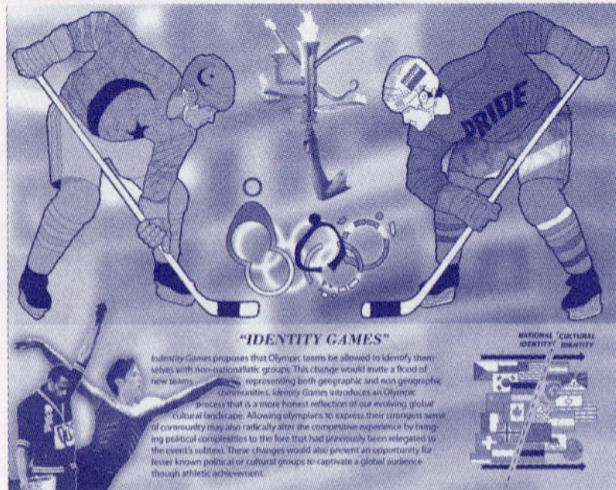

March 2010: The Olympics

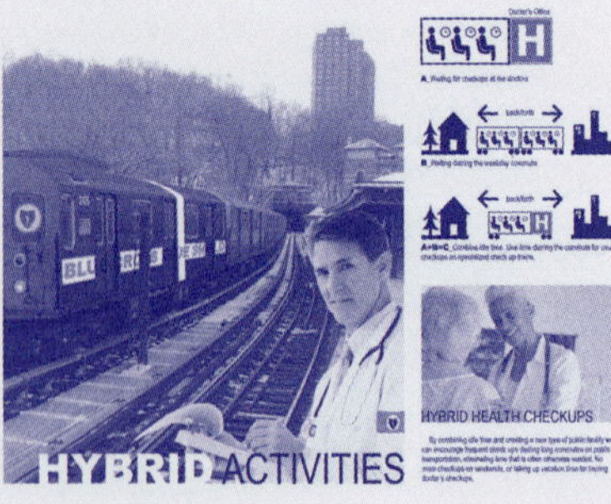

April 2010: Health Care

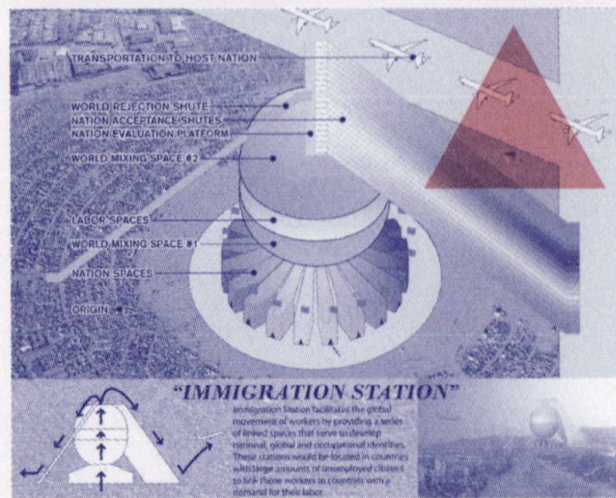

May 2010: Immigration

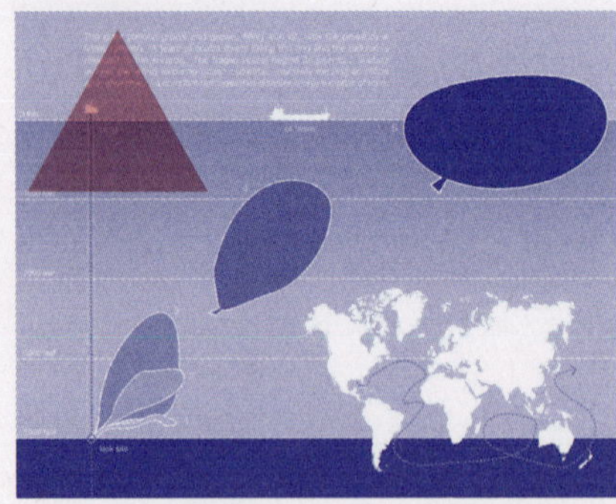

June 2010: Oil Spill

Neighborhood Watch: Joshua
Prince Ramus
Joshua Prince-Ramus

Soviet Contamination
Photographs and captions by
 Yevgeniy Fiks
Komar & Melamid, *Post-Art #1*
 (A. Warhol), from *Pictures of
 the Future* series, 1973–74.
Komar & Melamid, *Guggenheim
 Museum,* from *Pictures of the
 Future* series, 1973–74.
Former Komar & Melamid's Art
 Studio Archive.

Night Haunts musical score
Andrew Ingkavet

A Few Zines
Mimi Zeiger
Models: Gavin Browning,
 Kim Sulik and Carlos Solis

Rapid Response: Addressing the
Address
William Brian Smith

J. MAYER H.
© Dimitrios Tsatsas, Stylepark

Neighborhood Watch: Writers'
Block
Gavin Browning

The Commons
Gavin Browning: *The Commons,* directed
 and co-produced by Laura Hanna.
Written and co-produced by Gavin
 Browning.
Animated by Dana Schechter and
 Molly Schwartz.

Nora Libertun de Duren
New York City Department of Parks
and Recreation, Matthew Jamieson,
Steven Ko, and Nora Libertun de Duren.

Olympia Kazi
Olympia Kazi

Michael Mandiberg
Michael Mandiberg

Mark Shepard
Mark Shepard

Brooke Singer
Brooke Singer

The Geography of Buzz
Minna Ninova and Sarah Williams,
 Spatial Information Design Lab,
 Columbia GSAPP

ECOGRAM
Mitchell Joachim

Examined Life with Astra Taylor
and Avital Ronell
Alan Tansey

IRON DESIGNER
Ho Kyung Lee

Safari 7 Reading Room
Glen Cummings, MTWTF; Janette Kim
and Kate Orff, Urban Landscape Lab,
Columbia GSAPP, Gena Wirth, Lisa Ekle,
Sayli Korgaonkar, Jonathan Pettibone,
Robin Fitzgerald-Green, Evan Sharp

Imagining Recovery
Wayne Congar and Troy Therrien

InDisposed
Ben Ritter

Public Housing
GRITtv with Laura Flanders

*Red Lines Housing Crisis Learning
Center* Report at Studio-X
Damon Rich

Dispatches from Villa Feuerloscher
Hannes Preisch

ink
Michelle Fornabai

Ink Workshop with Karen Finley and
Michelle Fornabai, and page 17
Isabelle Rijnties

Reports from Rotterdam
Tobias Armborst, Daniel D'Oca,
Georgeen Theodore, Interboro

TRANS SIBERIA
Greta Hansen and Cheryl Wing-Zi Wong,
Warm Engine

SUPERFRONT at Studio-X
Kimberley Cases, set designer
Mike Deriex, drafter and builder
Mitch McEwen, unlicensed architect
Sarah Millsaps, structural consultant
Pawel Niedzwiecki, designer and builder

Dress Local
Sarah Cloonan, Federica Soletta,
Pollyanna Rhee and Tong Tong

Discussions on Networked Publics
Moro Arkaki, Alexis Burson, Christiana
Ciardullo, Leigha Dennis, Kyle
Hovenkotter and Kazys Varnelis

Spontaneous Architecture
January 2010: Daniel Georges
February 2010: Trevor Watson
March 2010: Andrew Miller and
 David Ruperti
April 2010: Trevor Watson
May 2010: Andrew Miller and
David Ruperti
June 2010: Adam Lemire

Index.

08.14.08
Neighborhood Watch: REX
Joshua Prince-Ramus, REX

08.23.08
Rapid Response: Beijing Air
Sarah Williams, Columbia
    GSAPP
Response by David Benjamin,
    Columbia GSAPP

09.23.08
Neighborhood Watch: 2x4
Michael Rock, 2x4
Georgianna Stout, 2x4

09.30.08
Rapid Response: Public Space
Eileen Clancy, I-Witness Video

10.14.08
Inside the Pavilion
Aaron Levy, Slought Foundation
Andrew Sturm, PARC
    Foundation
Laura Hanna, filmmaker

10.16.08
Live from Palestine
Raja Shehadeh, writer and
    lawyer

10.23.08
Neighborhood Watch: Toshiko
Mori
Toshiko Mori Architect

10.28.08
Rapid Response: Collapse!
Kazys Varnelis, Columbia GSAPP
Micah Fink, filmmaker
Daniel Buenza, Columbia
    School of Business
[co-sponsored by the Network
Architecture Lab, Columbia
GSAPP]

10.20.08, 10.27.08, 11.03.08,
    11.10.08
Reading Group: *Flying Close
to the Sun: My Life and Times as
a Weatherman*
Gavin Browning, Cristina
Goberna, Matthias Neumann,
Minna Ninova, Sierra
Feldner-Shaw, Abbot Street,
Cathy Wilkerson

11.18.08
re:construction
Daniel Perlin, artist

11.25.08
Rapid Response: I Need My
Space
Karen Finley, artist

12.02.08
Soviet Contamination
Barry Bergdoll, Columbia and
    MoMA
Jean-Louis Cohen, NYU
Yevgeniy Fiks, artist
Vitaly Komar, artist
Moderated by Jorge Otero-
Pailos, Columbia GSAPP

12.04.08
Night Haunts
Sukhdev Sandhu, NYU and *Daily
    Telegraph*
Andrew Ingkavet, composer

12.09.08
*Volume 17: Content
Management*, Launch Party
Jeffrey Inaba, Columbia GSAPP
Dana Karwas, Columbia GSAPP
Mark Wigley, Columbia GSAPP

01.08.09–02.28.09
*A Few Zines: Dispatches from
the Edge of Architectural
Production*
[curated by Mimi Zeiger]
Featuring: *Citizens of No Place,
Dodge City Journal, face b, The
Gospel of Judas, Infiltration,
junk jet, Lackluster, loud paper,
Monorail, PIN–UP, Shadows and
Straws, Situated Technologies
Pamphlets, Sumoscraper, Thumb*
Opening night panel:
Luke Bulman, *Thumb*
Felix Burrichter, *PIN–UP*
Stephen Duncombe, NYU
Mark Shepard, University at
    Buffalo
Mimi Zeiger, *loud paper*
Moderated by Kazys Varnelis,
Columbia GSAPP

01.27.09
Rapid Response: Addressing
the Address
Marisa Jahn, artist

Reinhold Martin, Columbia
    GSAPP
Andrew Ross, NYU
Dorian Warren, Columbia SIPA
Moderated by Mabel Wilson,
Columbia GSAPP

01.30.09
J. MAYER H.
Jeffrey Inaba, Columbia GSAPP
Andres Lepik, MoMA
Jürgen Mayer H., J. MAYER H.
Cristina Steingräber, Hatje
    Cantz Verlag
[co-sponsored by C-Lab,
Columbia GSAPP]

02.17.09
EVERYDAY THE URGE
GROWS STRONGER TO GET
HOLD OF AN OBJECT AT
VERY CLOSE RANGE
BY WAY OF ITS LIKENESS,
ITS REPRODUCTION
David Reinfurt, Columbia
    GSAPP
Lars Fischer, common room

02.19.09
Neighborhood Watch: Writers'
Block
ARO, ARROJO Studio, C-Lab,
Living Architecture Lab, Netlab,
Phaidon Press, REX, *Urban
China*, Waterworks and others.

02.24.09
Rapid Response: Towering
Inferno (TVCC)
Joseph Grima, Storefront for Art
    and Architecture
Jeffrey Johnson, Columbia
    GSAPP
Jyoti Hosagrahar, Columbia
    GSAPP and SIPA

02.26.09
Studio-X at the New Museum
David Benjamin, Columbia
    GSAPP
Sarah Williams, Columbia
    GSAPP
[co-sponsored by the New
Museum]

03.03.09
spurse: Deep Time + Rapid Time
Petia Morozov, Columbia GSAPP

and Soo-in Yang, Columbia
GSAPP
[co-sponsored by the Living
Architecture Lab, Columbia
GSAPP]

09.17.09
*Red Lines Housing Crisis Learning
Center* Report at Studio-X
Larissa Harris, Queens Museum
    of Art
Prerana Reddy, Queens Museum
    of Art
Damon Rich, Center for Urban
    Pedagogy
[co-sponsored by the Queens
Museum of Art]

09.19.09
Volume 20: Storytelling Launch
Party
Jeffrey Inaba, Columbia GSAPP
Katharine Meagher, Columbia
    GSAPP
Mark Wigley, Columbia GSAPP

09.22.09
Brydcliffe Report
Linda Weintraub, curator
Byron Bell, Bell Larson
Architects and Planners
Matthew Bialecki, Bialecki
    Architects
Solange Fabião, architect
Todd Rader, Rader + Crews

10.08.09
PRE-Office Conversations with
Architects
Zachary Colbert, Aaron Davis,
    Daniel Kidd and Leah
    Meisterlin, PRE-Office
David Benjamin, Columbia
    GSAPP
Laurie Hawkinson, Columbia
    GSAPP
Karla Rothstein, Columbia
    GSAPP
Soo-in Yang, Columbia GSAPP
[co-sponsored by KPFF
Consulting Engineers]

10.15.09–12.31.09
Safari 7 Reading Room
Glen Cummings, MTWTF
Janette Kim, Columbia GSAPP
Kate Orff, Columbia GSAPP
[co-sponsored by CeX Complete

Entertainment Exchange,
Columbia GSAPP and the Urban
Landscape Lab]

10.24.09
Landscapes of Quarantine Crit
Geoff Manaugh, *BLDGBLOG*
Nicola Twilley, *Edible Geography*
Guest critic: Bjarke Ingels, BIG

11.05.09
UiWE: Cultural Designers
Jacob Blak, UiWE
Christian Pagh, UiWE
Response by Dominic and
Christopher Leong, Leong
Leong Architects

11.10.09
Dispatches from Villa
Feuerlöscher
Barbara Hirnthaler, Villa
    Feuerlöscher
Gabriel Hirnthaler, Villa
    Feuerlöscher
Karen Finley, artist
Hannes Preisch, curator
Response by Mabel Wilson,
    Columbia GSAPP

11.17.09
The Garment District
Magda Aboulfadl, Manhattan
    Community Board 5
Barry Dinerstein, NYC
    Department of City Planning
Stan Herman, Council of
    Fashion Designers of
    America
Patrick Murphy, NYC Economic
    Development Corporation
Barbara Randall, Fashion Center
Business Improvement District
Moderated by Vishaan
Chakrabarti, Columbia GSAPP

11.18.09
Safari 7 Reading Room
Educators' Roundtable
Heather Cardinale, NYC
    Department of Education
Rachel Crumpler, Queens
    Museum of Art
Glen Cummings, MTWTF
Nathaniel Curtis, iLAND Art
Amanda Dargan, City Lore
Edward Eckert, Growing Up
    Green Charter School

Christopher Kennedy,
    Straspore and Solar One
Janette Kim, Columbia GSAPP
Helen Kongsgaard, Urban
    Landscape Lab
Jonathan Payne, Columbia
    GSAPP
Miriam Walls, NYC Department
    of Education
[co-sponsored by the Urban
Landscape Lab, Columbia
GSAPP]

11.19.09
Heather Rowe and Michelle
Fornabai in Conversation
Heather Rowe, artist
Michelle Fornabai, Columbia,
    GSAPP
[co-sponsored by Columbia
School of the Arts]

12.04.09–12.05.09
Architecture in Public:
A Workshop
[Anna Kenoff, Reinhold Martin,
Leah Meisterlin]
Barry Bergdoll, Columbia and
    MoMA
Cynthia Davidson, *Log*
Sarah Herda, Graham
    Foundation for Advanced
    Studies in the Fine Arts
Jeffrey Inaba, Columbia
    GSAPP
Geoff Manaugh, *BLDGBLOG*
Brooke Hodge, curator
Sylvia Lavin, UCLA
Amanda Reeser Lawrence,
    Praxis
David van der Leer, Guggenheim
    Museum
Andres Lepik, MoMA
William Menking, *The
    Architect's Newspaper*
Nicolai Ouroussoff, *New York
    Times*
Benjamin Prosky, Columbia
    GSAPP
Lisa Rochon, *Toronto Globe and
    Mail*
William Saunders, *Harvard
    Design Magazine*
Ashley Schafer, *Praxis*
Henry Urbach, SFMOMA
Mark Wasiuta, Columbia GSAPP
Mirko Zardini, Canadian Centre
    for Architecture

Amanda McDonald Crowley,
Eyebeam Art + Technology
Center
Christina Ray, CHRISTINA RAY
Mark Shepard, University at
Buffalo
Robert Sumrell, AUDC
Tim Ventimiglia, Ralph
Applebaum Associates
Moderated by Kazys Varnelis,
Columbia GSAPP
[co-sponsored by The MIT Press
and Netlab, Columbia GSAPP]

03.26.10–04.16.10
TRANS SIBERIA
Greta Hansen, Warm Engine
Cheryl Wing Zi-Wong, Warm
Engine

03.30.10
Rapid Response: Spontaneous
Architecture
theme: The Olympics
winners: Andrew Miller and
David Ruperti, Brooklyn
[co-sponsored by PRE-Office and
GOOD Magazine]

04.01.10
SUPERFRONT at Studio-X
Mitch McEwen, SUPERFRONT
Jace Clayton (DJ/rupture)

04.13.10
Discussions on Networked
Publics #3: POLITICS
Stephen Graham, University of
Newcastle
Deborah Natsios, Cryptome
Enrique Ramirez, *a456*
Moderated by Kazys Varnelis,
Columbia GSAPP
[co-sponsored by Verso and
Netlab, Columbia GSAPP]

04.15.10
The Just Metropolis
Jack Jaskaran, NYPD
Suman Raghunathan,
Progressive States Network
Javier Valdes, Make The Road
New York
Anne Frederick, Hester Street
Collaborative
Moderated by Julie Behrens and
Kaja Kühl, Columbia GSAPP

04.22.10–05.06.10
Imagination Vessels
Studio Sangue Bom 2010: Keith
Kaseman, Raul Correa-Smith,
Steven Garcia; Yihong Deng,
Hajar Ebrahim Darbandi, Keith
Greenwald, Sara Gutierrez
Armesto, Tzu-Husan Hsu,
Analdur Schram, Lior Shlomo,
Irene Urmeneta, Diego Urrego,
Nicolas Weiss, Li Yang Wu,
Daniel Yep

04.27.10
Rapid Response: Spontaneous
Architecture
theme: Immigration
winners: Andrew Miller and
David Ruperti, Brooklyn
[co-sponsored by PRE-Office]

04.29.10
GSAPP Alumni Weekend
Mark Wigley, Columbia GSAPP

05.04.10
Discussion on Networked
Publics #4: INFRASTRUCTURE
David Benjamin, Columbia
GSAPP
Frank Pasquale, Seton Hall Law
School
Mason White, University of
Toronto
Molly Wright Steenson,
Princeton University
Moderated by Kazys Varnelis,
Columbia GSAPP
[co-sponsored by Netlab,
Columbia GSAPP]

05.13.10–06.16.10
*Dress Local: Applied Mapping
of the Garment District*
Sarah Cloonan, Pollyanna Rhee,
Federica Soletta, Tong Tong
[co-sponsored by CCCPArch and
The Fashion Center Business
Improvement District]

06.24.10
Discussion on Networked
Publics #5: PUBLISH
[co-sponsored by *Domus* and
Netlab, Columbia GSAPP]

In this time of unprecedented global transformation,
which has generated so many urgent challenges
but also whole new forms of creativity, architecture's
unique ability to address both the most direct practical
problems facing global society and the highest
ambitions for that society becomes all the more important. As the field devoted to representing the greatest
aspirations of society, architecture (including urban
design, urban planning, historic preservation, and real
estate development) is a key lens through which to
see, understand, and participate in our evolving world.
Architecture is a form of optimism. More than
simply supporting the basic rhythms of everyday life,
it tries to envision a better life, turning practical dil-
emmas into the most expressive opportunities,
whether at the scale of a vast city, a building, a single
interior, or a small piece of furniture.

Yet the world we serve is changing so rapidly that whole
new forms of creativity, expertise and responsibility are
needed. China, the Middle East, Eastern Europe,
Latin America, South Asia, and Africa are acting as the
key laboratories for the future of the built environment,
generating whole new ways of thinking and urgent
questions to address. The change is so fast, the scale so
large, and the cultural and historical questions so deep
that schools of architecture have to evolve. Almost
all urban transformations today involve complex
dynamic interplays between unique combinations of
global forces. In such a world, it is critical that we
learn to exchange ideas in new ways while at the same
time immersing ourselves in the local conditions,
knowledge, history, and expertise in each region to
develop a new level of debate and global responsibility.
While GSAPP has long been very active on the ground
in all these regions, a new level of engagement is
urgently needed. Schools need to become students.

In recent years, GSAPP has adopted the label Studio-X
to refer to its most advanced leadership laboratories
for the future of the built environment. These have
to evolve at the same rapid speed as the urban environment itself. The label tries to capture the sense that
we have to be ready to face many unknown questions
that will arise and need to be engaged urgently,
creatively, and responsibly with a range of different

partners. A Studio-X offers a protective space for the private and collegial exchange of ideas still in formation, as well as a public gallery/lecture space, a website and publication program for the exhibition, communication and discussion of the thoughts and designs that result from this exchange. Such laboratories will be located around the world in a dynamic interactive network dedicated equally to practical problems in cities and to emergent thinking.

The vision of the Studio-X Global Network is intended to establish a unique exchange of ideas and people among key regional leadership cities around the rapidly evolving globe, including Beijing, New York, Moscow, Amman, Rio-São Paolo, and Mumbai. The aim of this exchange is a global partnership able to offer support to the highest possible level of reflection on new realities and active, intelligent, and productive engagement with those realities. Typically located in the historic downtown of these global cities, each Studio-X acts as an open platform for collaborative research and debate, along with a publication gallery, an exhibition gallery, a lecture space and an open studio workspace. During the day, the Studio-X is an active workshop, with combinations of ever-shifting teams of local experts and visitors from the region or globe working on designs, reports, exhibitions, books, competitions, films, magazines, etc. During the evening, the Studio-X acts as a hub of social exchange and intense debate with a lively program of exhibitions and events. It is a hot spot in the city, buzzing with social energy, invention, and dedication to a better future.

Each Studio-X is electronically linked in real time to every other Studio-X around the world, and ideas, people, and projects are continuously shared among them. The global Studio-X platforms are deeply integrated into the curriculum and research structure of all the programs at GSAPP, with students and teachers having multiple opportunities to spend time working in or with any Studio-X in the Global Network. Equally, students, teachers and experts from each platform around the world can work in or with any of the other platforms. The traditional hierarchical model of a leadership school concentrating expertise in a single place, synthesizing it and transmitting a singular approach to the major questions facing us, gives way to a distributed horizontal network model that can incubate evolving forms of intelligence for a new evolving world.

It is the ambition of the GSAPP to establish the most decisive global network of teaching, research and communication about the built environment. Such a global think tank must be based on the deep conviction that those parts of the world that are changing the most have the most to teach us. Older centers of power and wisdom must learn from the newer ones, which in turn learn from each other. No one city or region has a monopoly on the wisdom our shared world needs. Future thinking must be collaborative. Architecture can act as the key lens on our world, rather than the usual global frameworks (economy, public policy, ecology, etc.) in order to leverage the inherent optimism of the field into new kinds of visionary and practical understanding. The ability of architecture to reflect, magnify, communicate, and celebrate our highest aspirations must be turned into a powerful global tool.

The Studio-X Global Network is a massive undertaking that will take some years to complete. After the opening of the pilot Studio-X in downtown Manhattan in 2008, Studio-X Beijing and the Amman Lab were launched in March 2009, and have already become lively engaged sites. The Studio-X spaces in Brazil, India, Russia and Africa are currently being set up. With the addition of each hub in the network, this radical experiment in redefining the role, responsibility, and capacity of globally collaborative modes of education, research and action increases its bandwidth exponentially. A new kind of collective brain is emerging.

—Mark Wigley

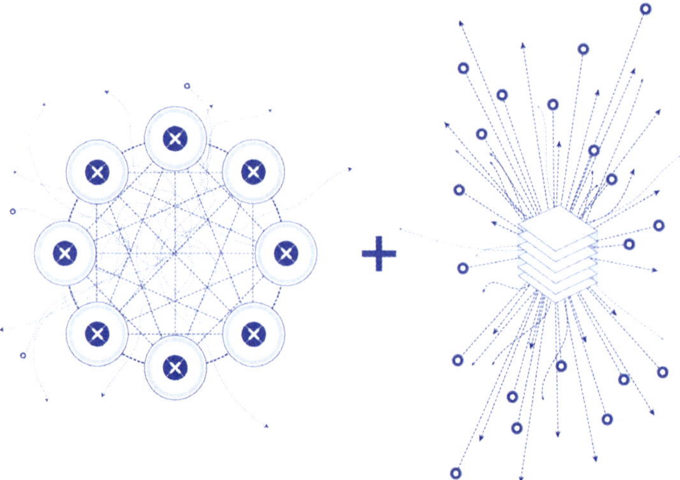

The Studio-X Global Network